Alisher Akramov

Justification of the main parameters of the seed dressing agent for pubescent sowing seeds

Alisher Akramov

Justification of the main parameters of the seed dressing agent for pubescent sowing seeds

Monograph

ScienciaScripts

Imprint
Any brand names and product names mentioned in this book are subject to trademark, brand or patent protection and are trademarks or registered trademarks of their respective holders. The use of brand names, product names, common names, trade names, product descriptions etc. even without a particular marking in this work is in no way to be construed to mean that such names may be regarded as unrestricted in respect of trademark and brand protection legislation and could thus be used by anyone.

Cover image: www.ingimage.com

This book is a translation from the original published under ISBN 978-620-8-11741-2.

Publisher:
Sciencia Scripts
is a trademark of
Dodo Books Indian Ocean Ltd. and OmniScriptum S.R.L publishing group

120 High Road, East Finchley, London, N2 9ED, United Kingdom
Str. Armeneasca 28/1, office 1, Chisinau MD-2012, Republic of Moldova, Europe
Printed at: see last page
ISBN: 978-620-8-26050-7

TABLE OF CONTENTS

INTRODUCTION

October 7 has been declared World Cotton Day by the UN General Assembly. In this regard, on October 7, 2019 in the city of Geneva at the held forum cotton is recognized for mankind as a Global Commodity. On the world cotton market the increase of confrontation for the market, on cotton-growing countries cultivation of new selections of cotton and zoning, on the basis of improvement of technologies of preparation of sowing seeds improvement of qualities and reduction of production costs and wholesale cost of seeds are actual tasks.

From the above described improvement of sowing qualities and reduction of seed cost in the world market, increase of germination with dressing, increase of resistance to diseases in all stages of seed preparation, including the use of chemical preparations, determination of factors negatively affecting the quality and their prevention, creation of resource-saving technologies reducing the cost of seed preparation remains the most prestigious task.

In the subglacial years in the republic complex measures are taken on cotton growing, development of its primary processing, expansion of assortment and types of finished products, including support of export opportunities of the enterprises of the branch.

Cotton-textile clusters are being created in the industry, cotton seed production, cotton growing and processing with the production of finished knitted fiber products are being developed, and measures are being taken to develop the system of export of finished products.

Having in mind the dependence of product quality it can be seen that the primary factor in the manufacture of quality textile products is the development of technology and techniques for the preparation of cotton seeds [1].

From the foregoing, the creation of technology for cotton seed treatment, providing uniform dressing with the consumption of a normalized amount of working suspension, improving the technology with resource-saving technical solution is an urgent task.

According to the Presidential Decree UP 5708 of April 17, 2019 "On measures to improve the system of state management in the sphere of agriculture" the main tasks of the Ministry of Agriculture of the Republic of Uzbekistan are-implementation of a unified state policy in the sphere of agriculture and food security, providing primarily for digitalization of agriculture, introduction of market principles in relations with the subjects of the agricultural sector, introduction of advanced experience and achievements of science, modern resource-saving and intensive agro-rotation

On the basis of the above it is possible to assert that development of a dressing agent for downy cotton seeds, allowing to increase uniformity and completeness of dressing is an actual task.

This study, to a certain extent, serves the fulfillment of the tasks stipulated by the decrees of the President of the Republic of Uzbekistan dated February 7, 2017 PF-4327 "Strategy of actions on five priority directions of development of the Republic of Uzbekistan in 2017-2021", dated December 14, 2017 PF-5285 "On measures on accelerated development of the textile and garment and knitwear industry", of November 28, 2019 PP-3408 "On measures to radically improve the management system of the cotton industry", as well as in other regulatory documents adopted in this area.

I. STATE OF THE ART AND OBJECTIVES OF THE STUDY

1.1. The role of pre-sowing seed treatment.

Formation of a unified policy aimed at increasing crop yields should be based on reliable protection of plants from pests, diseases and weeds through the use of highly effective low-toxic, environmentally friendly chemical and biological means [3].

The most widespread and harmful diseases of cotton include gommosis and root rots caused by a complex of phytopathogens. Gommosis and root rots cause a decrease in germination and plant stand density [4]. In cotton regions of Uzbekistan, gommosis occurs everywhere. To varying degrees, this disease is also observed annually in Armenia, Azerbaijan, Turkmenistan, Tajikistan, Kyrgyzstan, and southern Kazakhstan [5, 6].

Modern pre-sowing treatment of seed (planting) material is more complex than dressing, because during this process, besides fungicidal or insecto-fungicidal dressing agents, seed (planting) material is also treated with protective and stimulating materials containing plant growth stimulants, complex microfertilizers, individual microelements and film-forming substances that provide both protection and stimulation of plant growth processes [7].

Proper pre-sowing seed treatment increases their field germination, reduces the infestation of plants by pests, diseases and allows to obtain a given number of seedlings without thinning [8, 9, 10, 11, 12, 13].

The use of untreated seeds for sowing causes a decrease in crop yield. To limit the harmfulness of viral diseases of wheat, barley and oats and damage by their pests (cereal aphids), it is more reliable, economically more profitable and environmentally safer to conduct annual pre-sowing disinfestation of seeds than to use repeated spraying of crops with contact (pre-harvest) and systemic (organophosphorus, carbonate) insecticides, because aphids quickly develop resistance to the latter, so the effectiveness of treatment is significantly reduced [14].

Crop yield losses from diseases, weeds and pests can reach more than 30%, of which 14% from pests, 12% from diseases and 9% from weeds [15].

Long-term practice and world experience, as noted above, proves that about 30% of the saved yield depends on qualitatively conducted seed dressing, correctly selected dressing agent [16].

Not so long ago, seed treatment with chemical preparations was necessary to control root rot, bunt, snow mold, septoriosis and powdery mildew in the early stages of crop development [16].

Root rot, a very common disease, is found in all cotton-growing areas of Uzbekistan. It is also widespread in many other countries, such as Russia, a number of countries in Southeast Asia, China, Egypt, USA, Mexico, etc. [17, 18, 19, 20, 21]. [17, 18, 19, 20, 21].

Root rots of cotton should be considered as complex diseases. The cause of their occurrence is not only soil microorganisms, but also unfavorable environmental conditions during the period of plant growth and development, as well as the quality of soil tillage before sowing, seedling care, fertilizer application, irrigation and other factors that lead to changes in temperature, air and water regimes of the soil and thus increase or decrease plant resistance to the disease. Of particular importance in increasing the resistance of seedlings to root rot are the sowing qualities of seeds [22].

Root rot of cotton seedlings is a very widespread disease. The causative agent of the disease is a complex of microorganisms living in the soil - mainly the fungus Rhizoctonia solani, as well as some species of Fusarium and others. They can develop both on living and dead plant parts. These microorganisms cause rotting of seeds, seedlings, sprouts and even young plants at the age of the first two true leaves [23].

It was determined that the root rot pathogen also affects seeds. Seeds affected by gummosis are underdeveloped, remain puny and often lose germination [19]. Therefore, for better protection against diseases, cotton seeds should be dressed in preparation for sowing [19, 24, 25].

A common feature of the above cotton diseases is seed transmission. However, the importance of this mode is not the same for each disease. For wilt and homosis, seed is the main source of transmission from year to year. Root rot infections are mainly found in the injured seed coat and in the soil, and seedlings are affected through it [17].

Pre-sowing seed treatment to protect it from pests and diseases is one of the necessary and effective measures [26, 27] and is carried out for [28, 29, 30, 31, 31, 32, 33, 33, 34, 35, 36]:

- seed disinfection against seed-borne pathogens;

- protecting sown seeds from mold diseases during germination.

Mordanting provides an opportunity:

- disinfect seeds from plant pathogens that are transmitted through seed;

- protect seeds and seedlings from damage by phytopathogenic organisms;

- reduce damage to seedlings by root grills, as well as to soils by dwelling and terrestrial pests at the initial stage of plant development;

- reduce the negative impact of traumatic damage to seeds as a result of activation of its protective properties and prevention of microorganisms development;

- stimulate growth and development, increase resistance to unfavorable external factors of seeds and plants due to the effect of preparations on some physiological processes of germinated seeds and plants;

- increase the quantity (up to 5 c/ha on average) and significantly improve the quality of yield [16].

In this regard, seed pre-sowing treatment is one of the main operations in the complex of measures to control pests and diseases of agricultural crops in industrialized countries [15, 37, 38]. According to experts, cotton pests and diseases cause yield losses of up to 20-30% [19, 39, 40].

In North America, Vitavax seed treatment increases wheat yields by 7.9% and barley yields by 10.8% by increasing germination rates, the number of ears per 1 m^2 and grain weight [41].

In general, according to experts, reliable protection of cotton against diseases can be realized only by using all existing measures proposed by scientific institutions and best practices [42].

Seed dressing of agricultural crops provides an increase in their yield, quality and production efficiency [43, 44, 45, 46, 47, 48, 49]. Justified choice of dressing agent and technical means of its application on seeds and quality treatment of seeds before their dressing are one of the first conditions for effective crop production [49, 50, 51].

Costs for pre-sowing treatment are recouped quickly enough in 7.4...14.6 times, and in case of ultra low-volume treatment at insignificant consumption of chemical preparation in 70...400 times in comparison with untreated seed [52].

Seeds to be dressed shall conform to the state standard [O'z DSt 663:2017] Table 1.1.

Table 1.1.

Qualitative indicators of sown seeds

Name of indicators	Norm, %		
	For downed seeds	For seeds of low pubescenc e	For bare seeds
Germination, not less	90,0	90,0	90,0
Humidity (mass fraction of moisture), not more than	10,0	10,0	10,0
Contamination (mass fraction of mineral and organic sap), no more than	0,7	0,5	0,3
Mechanical damage, not more	7,0	8,0	8,0
Droopiness, not more	-	2,5	0,5
Residual fibrousness, not more: - downy - for the naturally bare	 0,8 0,4		

Powdered, paste-like and liquid water-soluble preparations, their consumption rates and dressing periods should also comply with the standards. Dressing is carried out at positive air temperatures in specialized seed preparation shops [53], or in open flat areas, under sheds or in well-ventilated rooms [54]. Quality indicators of the technological process [16, 53, 55, 56, 57]:

Completeness of dressing, %	100 ± 20
Unevenness of seed and working fluid supply to the treater, %	± 5
Unevenness of suspension concentration, %	± 5
Irregularity of dressing agent supply, %	±5

1.2. Methods of seed dressing of agricultural crops

Analysis of literature, scientific research, as well as production experience show that currently there are the following methods of dressing in production [14, 44, 51, 55, 56, 57, 58, 59, 60, 61] (Fig.1.1):

- moisturized dressing;
- semi-dry dressing;
- wet dressing
- dry dressing;
- fine dressing;
- thermal treatment method.

The most widely used method of seed pre-sowing treatment is chemical dressing. This method emerged at the turn of the 20th century and is actively used nowadays [62]. As a rule, in Russia, about 60% of seeds are dressed in large and medium-sized farms and 40% in farms [63, 64, 65, 66].

The dry dressing method consists of coating (powdering) or mixing the seeds with a dusty pesticide. Compared to other methods, seeds can be dressed long before sowing with the least amount of chemical. Seeds treated in this way are well preserved and do not require additional treatments. But the preparation is poorly retained on the surface of the seeds, and due to spraying of the pesticide, sanitary and hygienic working conditions of workers deteriorate and part of it is lost.

Semi-dry method consists in treatment of seeds with atomized suspensions, stronger solutions of e.g. formalin and keeping seeds in piles for 2...4 hours. This method provides high uniformity of coverage of seeds with pesticide and creates better sanitary-hygienic conditions for workers. Its disadvantages include low productivity and excessive moistening of seeds.

Thermal method of treatment is used to control pathogens of cotton seeds by immersing seeds in water heated up to 50^0 C. The moisture content of seeds is increased by 10.15 %. At this method of dressing the moisture content of seeds is increased by 10...15 %, after which the seeds must be dried to a conditioned moisture content. Thermal method of treatment should be carried out with precise observance of the regime, because overexpression of seeds in hot water leads to a sharp decrease in seed germination, and underexpression does not kill the infectious beginning [67, 68].

The wet method consists in wetting seeds with a solution of dressing agent. Wetted seeds for 2...3 hours. They are kept under a tarpaulin, then dried. Wet dressing is carried out 2...3 days before sowing. The method is low-productive, very labor-intensive, as it requires considerable labor costs for drying after its dressing.

Another type of pre-sowing seed treatment technology is wetting dressing, when suspensions, solutions, powder preparations are applied to the seed surface with simultaneous or subsequent wetting with liquid at the rate of 10-30 liters per ton. This method of dressing has a lot of undeniable advantages, notes Mozharova. For example, economical use of the drug due to the accuracy of dosing the liquid, good quality of treatment, the ability to apply simultaneously with pesticides, micro- and macro-fertilizers, a small wetting of seeds and no need for their subsequent drying, as well as satisfactory sanitary and hygienic working conditions. The process can be mechanized with high technical and economic indicators. The disadvantage of the method is shattering of dressing agent from seeds as it dries, adhesives are used for better retention of preparations, unfavorable sanitary and hygienic working conditions and environmental pollution [2, 60].

The fine dressing method consists in treating seeds with a suspension - a mechanical mixture of atomized dressing agent with water, in which the smallest particles of the dressing agent are in suspension.

The suspensions are fed under high pressure through small atomizer openings. Due to this, the rate of drug consumption is reduced, the quality of dressing is increased and the increase in seed moisture content is not more than 1%. Therefore, seeds dressed by this method do not require additional drying and can be stored for a long time before sowing [14, 44].

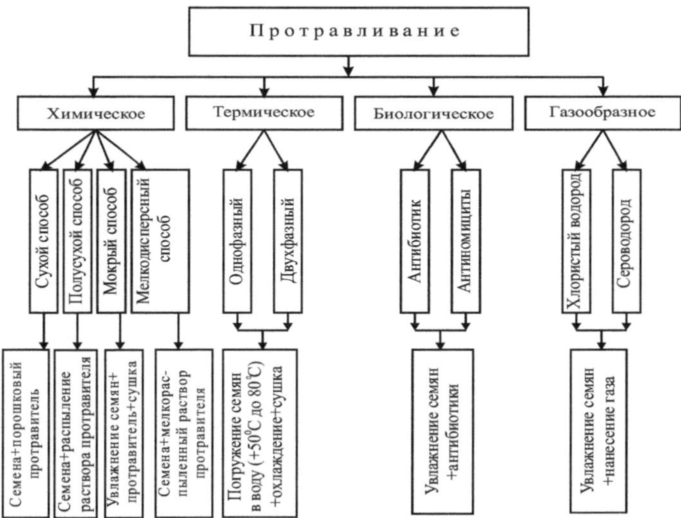

Fig. 1.1. Classification of technologies of seed pre-sowing treatment

The main requirement for dressing is to ensure the high quality of the process itself in order to realize the full effectiveness of the product. For this purpose, small amounts of dressing agent should be applied evenly to the seeds. For quality dressing it is necessary to use thoroughly cleaned seeds [14].

Thus, the considered methods of dressing and agronomic requirements for them show that there are many methods of dressing, however, they have their advantages and some disadvantages, so it is necessary to work for the development of specifically selected method of dressing.

1.3 Classification of dressing agent spraying methods and types of sprayers

Atomization is the splitting of a jet or film of liquid into a large number of droplets and their distribution in space (volume of chemical-technical apparatus). Devices for atomization, equipped with one or more nozzle openings, are called atomizers or nozzles, and the flow of droplets - spray. The methods of atomization are extremely diverse [69].

Sprayers used in practice mostly disperse the working fluid with the drug into droplets of various sizes, i.e. form polydisperse systems. The same applies to powdery, dust-like preparations [70]. In the best case it is possible to regulate the average particle size, and large weight fractions of different-sized particles reduce the efficiency of treatment and adversely affect the uniformity and coverage area. In the literature, researchers cite various classifications of methods of atomization of liquids. However, according to the classification of L.A. Vitman, B.D. Katsnelson and I.I. Paleev, two large groups of methods are emphasized - mechanical and pneumatic methods of atomization [71, 72] of liquids and powders (Figure 1.2).

Hydraulic atomization is determined by pressure of liquid supply (0,35-70 MPa). Advantages: in comparison with other methods, economical energy consumption (2-4 kW per 1 ton of liquid), simplicity and reliability of equipment; disadvantages: inhomogeneity of atomization, difficulty in regulating the flow of liquid at a given quality of crushing and dispersion of viscous liquids [69, 73].

The liquid (dressing agent) passes through the nozzle and due to the injection pressure, acquires a sufficiently high velocity, then transformed into a form that promotes its rapid atomization into individual droplets [69, 74, 75].

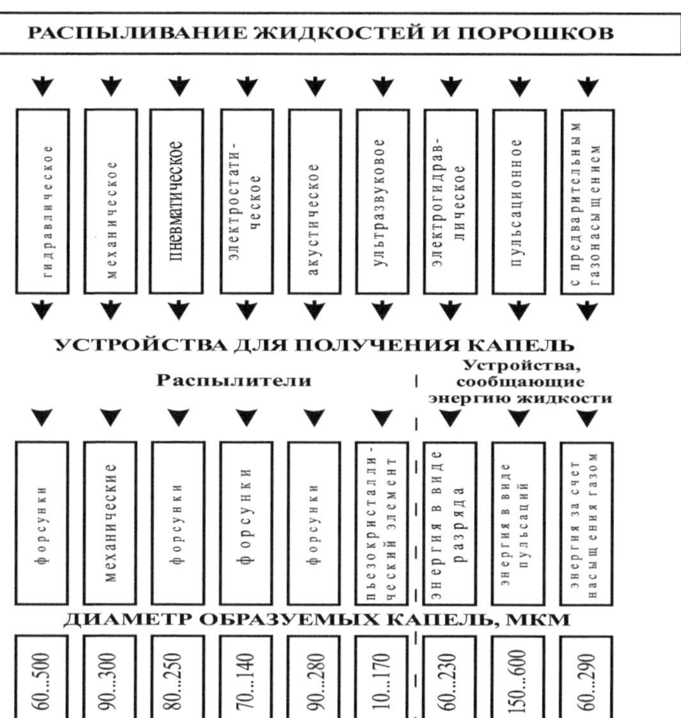

Figure 1.2. Classification of spraying methods and types of atomizers

In mechanical atomization, the liquid is energized by friction against a rapidly rotating working element (disk, cup, star, cone, nozzle and jet), acquiring rotational motion with it. Under the action of centrifugal forces, the liquid is blown off the atomizer (in the form of films or jets) and crushed into droplets. Advantages: the possibility of dispersing highly viscous and contaminated liquids and suspensions, regulation of their flow rate without changing the dispersibility, atomization at low flow rates of contaminated and highly viscous suspensions (liquids) into individual drops of the same adjustable size [76, 77], disadvantages: means energy intensity (15 kW per 1 ton of liquid), complexity of manufacturing and operation of atomizers. Distinguish mechanical atomizers of two groups: with direct fluid supply to the working element and immersed (Fig. 1.3 a, b). The first group includes atomizers with disk, cup, star, nozzle and jet elements, the second group includes disk and cone [69, 71] (Fig. 1.3 c-z).

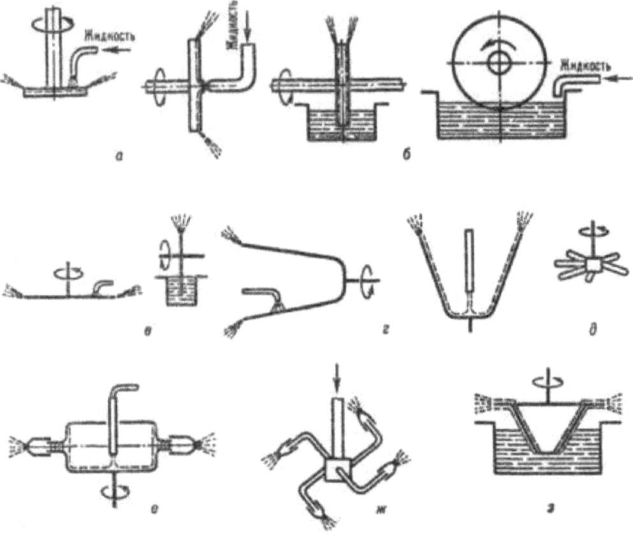

Figure 1.3. Mechanical atomizers

The flow of liquid through a mechanical atomizer with a horizontal axis of rotation (disk) and the formation of droplets behind the edge of the disk have been considered quite fully by Dunsky V.F., Nikitin N.V. [78, 76]. [78, 76].

Pneumatic sawing is caused by the interaction of. liquid with the sawing gas, as well as the resulting mixture with the environment. Advantages: low dependence of dispersing quality on liquid flow rate, reliability of atomizers, possibility of crushing high-viscosity liquids; disadvantages: high energy intensity (50-60 kW per 1 ton of liquid), the need for sawing gas and equipment for its supply. Nozzles of this type (Fig. 1.4) are divided into groups: by pressure drop, by the place of contact (external or internal mixing), by the character of flow movement (straight jet or vortex with swirling of gas or liquid), etc. [69, 75].

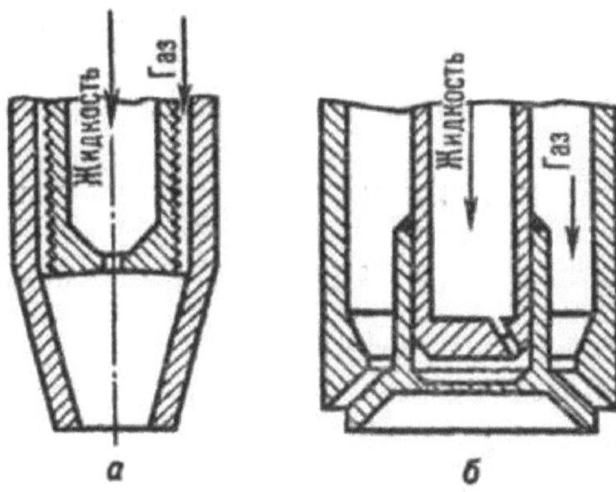

Fig. 1.4. Pneumatic nozzles: *a*- internal mixing with gas swirl, *b*- external mixing with liquid swirl.

In the electrostatic atomization method, the working fluid is fed into the electrostatic discharge zone, where it breaks up into individual droplets in the form of a film. Nozzles with artificial electrification of liquid atomization are divided into three groups: for application of various coatings, for atomization of liquids and powders. The disadvantages of this method are the need for expensive and complex equipment, its low productivity and rather high energy consumption [79].

Atomization of liquids by acoustic method is similar to pneumatic atomization in a number of ways; the liquid also receives energy due to the directional action of high-speed gas flow, but in this method the gas flow is communicated oscillations of different ultrasonic frequency. This provides a fine and homogeneous (monodisperse) crushing of the liquid into individual droplets. Acoustic nozzles are subdivided according to the type of acoustic vibrations generated by a special device and are classified into five nozzle groups and types of generator: with static; with vortex; with dynamic, etc. The acoustic method of atomization is considered to be more low-cost than the pneumatic method, but the main disadvantage is the complexity of the construction of these types of nozzles [69].

The ultrasonic method of atomizing a liquid allows atomizing it into fairly finely dispersed droplets. Atomization is carried out when the liquid is fed to the element of magnetostrictive or piezoelectric generator, which is in the mode of oscillation with ultrasonic frequency. Technical devices that atomize the liquid into droplets by ultrasonic method are small in capacity (0.5-6 kg/h), in addition, their design is difficult to manufacture and expensive [69].

The same disadvantages are noted in the pulsation method of atomization. Pulsating pressure flows act on the liquid flow, increasing the energy of the liquid film, and as a consequence, atomization into small droplets. This method is used in combination with various methods: mechanical, hydraulic, and pneumatic.

Unfortunately, as can be seen from the classification, these atomizers, except for mechanical ones, are not capable of creating monodisperse systems from viscous working solutions. Only some devices are known that allow to create drops of approximately the same size within 5-50 microns - capillaries, droppers, rotating perforated drums [76]. Unfortunately, their use in the design of technical devices for pre-sowing treatment is extremely difficult due to low productivity, inability to work with highly viscous liquids.

In the existing technologies agrotechnical requirements are made mainly for dressing and technical means of its realization. And even in them there are no modern regulated requirements for these types of devices. Therefore, at present it is necessary to develop both requirements for the use of environmentally safe technological methods, and to improve technological schemes of treatment and technical devices, and, first of all, as the most promising, used for treatment with biopreparations and microelements. For this purpose, it is necessary to reveal the directions of improvement of technical means of seed pre-sowing treatment.

1.4 Factors determining seed dressing quality

The quality of dressing depends on many factors related to the condition of the seed, the characteristic of the dressing agent and its prepropositive form, the construction and technological scheme of the dressing machine.

At all specialized seed preparation workshops of the Republic, quality control of dressing is carried out by visual comparison of dressed seed sample with a standard specially prepared for this purpose. Every year before the beginning of the season of preparation of sowing seeds for all shops is prepared (last year's standards are updated) standards of dressed sowing seeds. For this purpose, a small certain amount of seed prepared for dressing is taken from the seeds prepared for dressing and the necessary amount of dressing agent (working suspension) is weighed to be applied to the sample. Up to now this process has been done manually, i.e. after applying the dressing agent to the seeds, mixing is done manually.

To facilitate manual labor and mechanization of preparation of standards of dressed pubescent seeds, the author developed and manufactured a device for preparation of standards [80]. In the future, to verify the reliability of the results of experiments conducted in this work, the etalons of dressed pubescent seeds prepared by this device were used.

The main requirement for dressing is to ensure the high quality of the process itself in order to realize the full effectiveness of the product. For this purpose, small quantities of dressing agent must be applied evenly to the seeds. For quality dressing, thoroughly cleaned seeds must be used.

This additional cleaning economically and reliably prevents technical seeds from entering the dressing agent [14].

It should also be noted that quality dressing is only possible if it is carried out by qualified personnel with the right combination of all the above mentioned components. A high quality dressing can only be achieved if the following criteria are met [51, 61]:

- Firstly, compliance with the recommended consumption rates, i.e. the amount of dressing agent required for a certain volume of seed must be precisely maintained;

- Secondly, the preparation, respectively the active ingredient, must be evenly distributed over the entire surface of each individual grain;

- Thirdly, the adhesive used in the dressing must ensure that the entire dose of active ingredient applied to the grain is retained even after mechanical stresses such as storage, bagging, transportation and sowing;

- Fourthly, seed traumatizability, after dressing, should not exceed the norms of agrotechnical requirements.

The quality of dressing depends on many different factors. The analysis shows that the factors determining the quality of crop seed dressing can be grouped into four main groups (Fig.1.5):

- physical and mechanical properties of seeds;
- physicochemical properties of the dressing preparation;
- technological factors;
- factors depending on the design of the dressing machine.

Fig. 1.5. Structural diagram of factors influencing the quality of etching

The first group of factors includes the operating mode and technological adjustments of the dressing agent. Since the setting of the operating mode and technological adjustments of the treater is performed by the operating personnel, this group of factors can be referred to human factors. Therefore, persons who are familiar with the device, technological adjustments and operating modes of the dressing machine are allowed to work. Much depends on the qualification of the working personnel, on how they adjust the dressing machine to the operating modes, how they prepare the working solution.

Structural and technological scheme of the seed dresser, design of individual working bodies, material of working bodies and geometrical parameters of working bodies can be referred to constructive factors. Physical-mechanical properties of seeds: moisture, dustiness, size, bulk weight, size uniformity, hardness, weight per thousand seeds, ossification, etc., etc.

1.5 Status and development of methods and technical means for crop dressing.

For dry dressing B.N. Emelin offers a boom dressing machine (Fig.1.4). The technological process is carried out as follows: the operator periodically introduces the boom 8 into the grain mound. He opens the shut-off valve 6 on the boom at the beginning of its immersion into the bulk and closes it before removing the sawing tip.

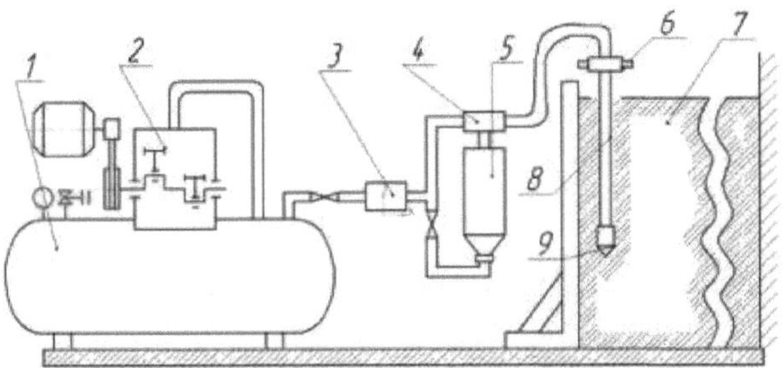

Fig. 1.6. Boom etchant B.N. Emelin: 1-receiver; 2-compressor; 3-water-oil separator; 4-reverse valve; 5-dosing feeder-feeder of pesticides; 6-shut-off valve; 7-template overlay; 8-boom; 9-rod.

Dry powders have the advantage that they are easy to use. Even in the simplest installations, e.g. drums or concrete mixers, a very good and uniform distribution on the grains is ensured. In addition, the seed can be processed regardless of the ambient temperature, even in severe frost [81].

In dry dressing, however, the deteriorated adhesion of the preparation has a negative effect. Under certain conditions, this can lead to dust emission at the place of work of the personnel and to significant losses of the preparation (up to 30%) [61].

As noted by I.P. Maslo [82], sanitary and hygienic labor conditions deteriorate. These disadvantages are reduced by moistening seeds and powder during dressing (Fig. 1.7).

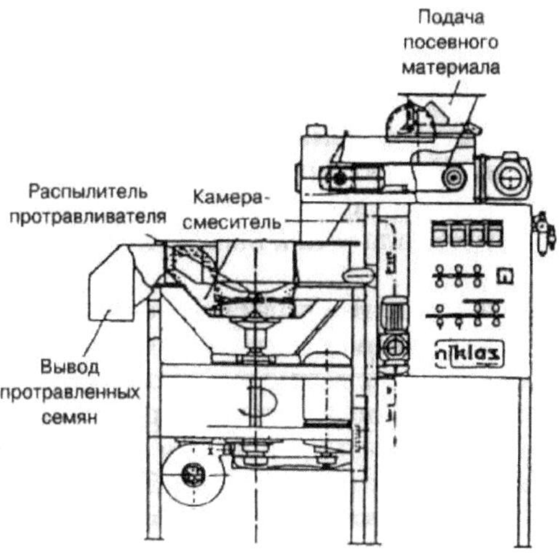

Fig. 1.7. Fully automatic stationary seed dressing machine for wet seed dressing and encrustation with electronic control (Niklas company)

Moisturizing dressing plants include a spray chamber through which the treated seed passes as a thin layer, which is sprayed with the required amount of dressing agent [61].

A special influence on the uniform distribution of the preparation on the seed has a special influence on the design of technical means in terms of the preparation and treatment chambers, as well as the type of device for feeding the preparation and seed into the treatment chamber. There are two main types of technical devices for treatment: flow type machine - the supply of preparation and seed material into the treatment chamber is continuous; batch type machine - a certain weight of seed material into the treatment chamber is supplied with a certain measured amount of preparation. It is considered that machines of flow type have a higher productivity, but less uniformity of drug deposition on the seed. In Western European countries all seed treatment plants are of stationary and inline type. Seed is fed by means of elevators or a separate loading auger [83, 84].

Modern production of seed dressing agents for agricultural crops at the present stage is carried out in the following directions [85, 59, 86].

a) Copying the design and operating principle of screw and chamber dressing machines serially produced in the past years, based on the use of modern components (pumps, pressure regulators, electrical devices, electronic automation systems, remote control) and materials (stainless steel, polymer products, etc.).

Among such technical means are mainly used high-performance mobile chamber grain dressers PK-20 "Super" ("Lvovagromashproekt", Ukraine), PS-5 "Farmer" (Belarus) (Figure 1.8), PS-10A ("Gatchinselmash", Russia) (Figure 1.7), with the help of which the dressed seeds are loaded into the autoloader of seed drills, and screw dressers like PNSh-5 and PNSh-3 ("Lvovagromashproekt") or Gramax-V (Farmgep) with packing of dressed seeds into bags.

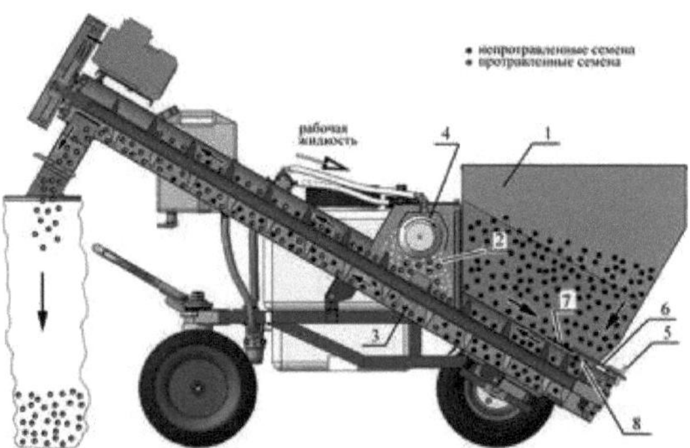

Figure 1.8. Technological scheme of work of dressing machine PS-5 "Farmer": 1-bunker; 2-chamber of dressing; 3-shaker blades; 4-nozzle; 5-adjusting flywheel; 6-adjusting screw; 7-grain doser flap; 8-window doser.

Table 1.2.

Technical characteristics of PS-5 "Farmer"

Capacity for 1 hour (on wheat), t/hour	0,7-3,4
Completeness of dressing, %	100±20
Mechanical damage of seeds, %, not more	0,5
Increase in seed moisture content, %, not more	1,0
Seed feeding irregularity, %, not more, not more	±5
Unevenness of working fluid supply, %, not more	±5
Power consumption, kW, not more	1,4
Overall dimensions in working position, mm	2300×1270×1470
Specific power consumption, not more, kWh/t	0,42

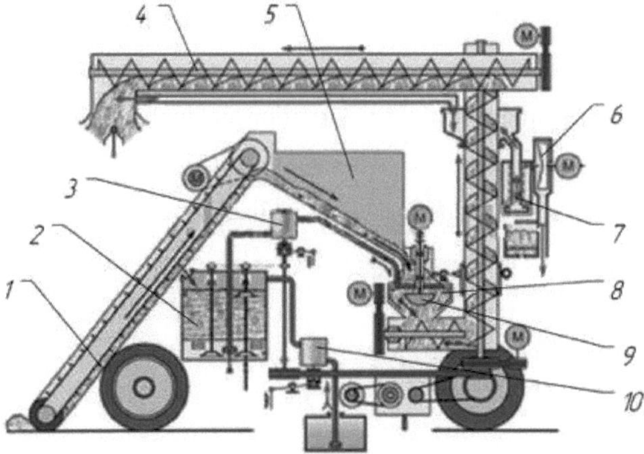

Fig.1.9 Scheme of seed dressing machine PS-10AM: 1 - loading conveyor; 2 - reservoir for working liquid; 3 - dosing unit; 4 - unloading auger; 5 - seed hopper; 6 - fan; 7 - filter; 8 - dressing chamber; 9 - spraying disk; 10 - pump.

Table 1.3.

Technical characteristics of PS-10AM

Productivity per hour of basic time, tons	up to 20
Tank capacity, l	200
Metering feed rate, l/min	0,5-3,5
Movement speed during maneuvering, m/sec	0,4
Power requirement, kW	5,6

b) Improvement of the design and technological scheme and the principle of operation of dressing machines serially produced in the past with the following changes:

- simplification of design and reduction of energy consumption by reducing technological processes (self-filling with water, electric heating of the working fluid, possibility of rotation and inclination of the discharge auger, exclusion of the system of suction and purification of poisoned air, etc.). Russian-made seed dressing machines PS-20M-4, PSS-20, UPS-10 can be mentioned as an example (Fig. 1.8);

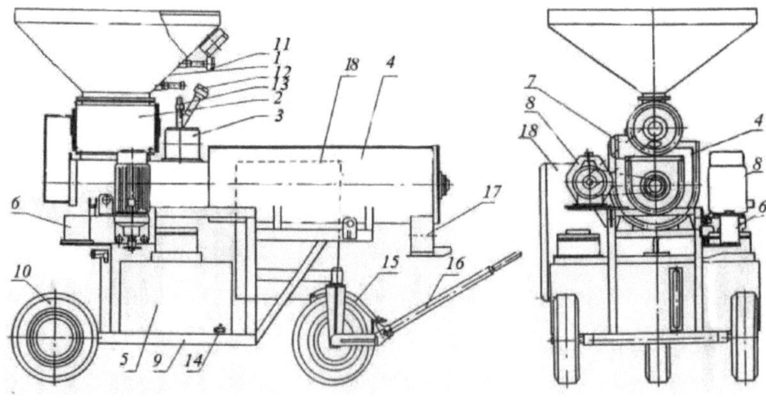

Figure 1.8 Schematic diagram of the installation of pre-sowing seed dressing UPS-10 1-receiving hopper; 2-dosing unit of seeds; 3-spraying chamber; 4-mixing chamber; 5-tank of working liquid; 6-pump-dosing unit; 7-spraying device; 8-electric drive; 9-frame; 10-wheel drive; 11, 12-level sensor; 13-supply pipe; 14-drain cock; 15-parking brake; 16-pull rod; 17-unloading device (bag holder); 18-control panel.

Table 1.4.

Technical specifications UPS-10

Wheat seed treatment capacity, t/h	10
Working fluid control range dispenser, l/min	0,5-3,5
Uneven feeding of the seed to be treated of material and working fluid, %	±5
Inhomogeneity of slurry components distribution over the volume of the working liquid container should not exceed in the process of operation, %	5
Seed crushing, %, not more	0,5
Overall dimensions, mm	1200×2900×2300
Weight, kg	850±50

- use of new technical solutions in the design of auger and chamber unit in order to improve the quality of application of working liquid on the surface of seeds: auger with the possibility of axial movement, turbo disk seed dispersers, cascade grids, disk (flat, wavy) or nozzle sprayers of working liquid. These include seed dressers PS-20 (Russia) (Fig. 1.9); ST 2-10/ ST 5-25 (firm "Petkus" Germany) (Fig. 1.10); screw seed dresser according to the patent of thc Russian Federation №2217897 (Fig. 1.11) [87]; chamber seed dresser according to the patent of the Russian Federation №2316925 (Fig. 1.12) [88];

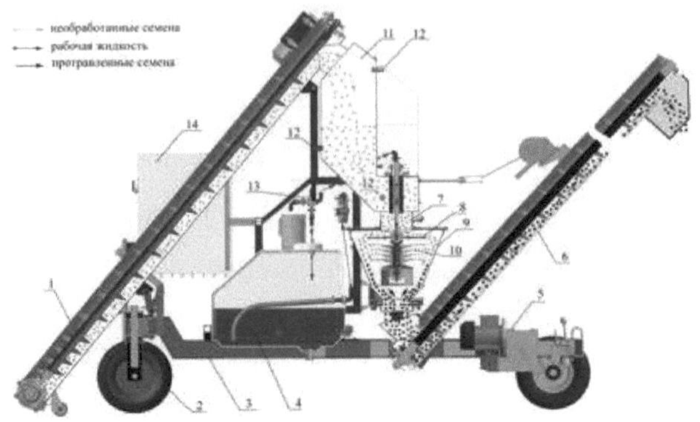

Fig. 1.9 Technological process of dressing machine PS-20

1-loading auger; 2-wheel; 3-frame; 4-tank for pesticides; 5-mechanism of self-moving; 6-unloading auger; 7-seed dispenser; 8-dispersing disk; 9-chamber dressing; 9-sprayer; 10-cascade disk; 11-bunker of seeds; 12-sensors; 13-dispenser of pesticides; 14-control panel.

Chamber seed dressing machine PS-20 (Fig.1.9) is an automatic self-moving machine with electric drive of the main mechanisms and consists of the following assembly units: hopper for seed accumulation, dressing chamber, tank for working liquid, dosing pump, unloading auger, loading auger, front wheels turning mechanism, self-propulsion, liquid flow control unit, control cabinet.

The supply of seed and working fluid to the dressing chamber is synchronized by three sensors that are mounted on the seed hopper [48, 54].

Table 1.5.

Technical characteristics of PS-20

Capacity (wheat), t/h	20
Completeness of dressing, %	100±20
Mechanical damage of seeds, %, not more	1,0

Increase in seed moisture content, %, not more	0,5
Seed feeding irregularity, %, not more, not more	±5
Unevenness of working fluid supply, %, not more	±5
Power consumption, kW	6
Weight, kg	750

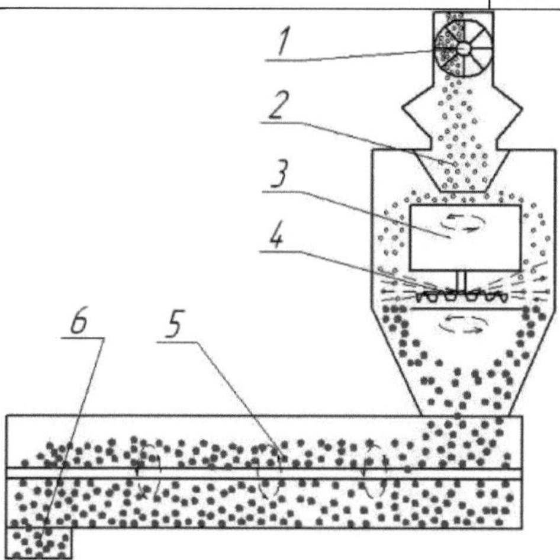

Fig. 1.10. Schematic diagram of dressing machine ST 2-10 PETKUS: 1-gate metering unit; 2-feeding hopper; 3-distributing disk; 4-sawing disk; 5-additional mixer; 6-outlet.

Table 1.6.

Technical data ST 2-10 PETKUS

Capacity (wheat), t/h	2-10
Installed engine power : Disk, kW Cleaning device (optional), kW Additional mixer, kW	 0,55 0,18 1,1
Sluice gate dispenser, kW	0,37

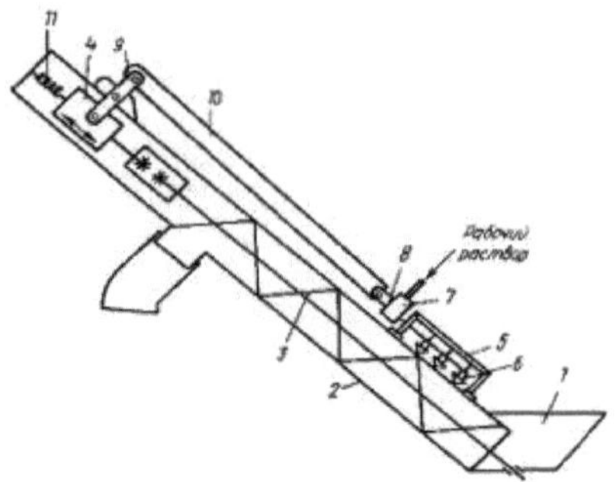

Fig. 1.11. Schematic diagram of the screw seed dressing machine, RF patent No. 2217897: 1-hopper; 2-hull; 3-screw; 4-sensor; 5-bar; 6-sprayer; 7-doser; 8-executive mechanism of the doser; 9-two-shoulder lever; 10-pull rod; 11-spring.

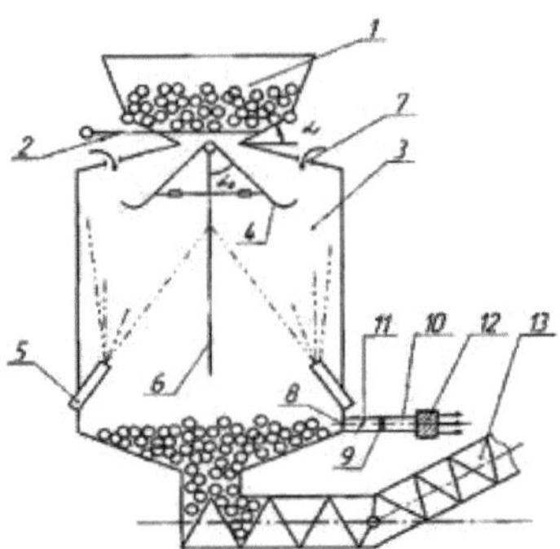

Fig. 1.12. Chamber seed dressing machine RF patent №2316925: 1-hopper; 2-flap; 3-chamber dressing; 4-distributor; 5-sprayer; 6-baffle; 7-ventilation hole; 8-suction pipe; 9-fan; 10-pipe; 11-flap; 12-filter; 13-screw.

Various modifications of dressing machines are used for dressing of cotton seeds in dressing shops of cotton plants and procurement points. Thus, for dressing of bare sown cotton seeds, dressing machines APH-5, 2-OSH, UOSH-6, as well as a set of equipment of KPS-15 brand and a similar set of KPS-19 brand are used, and for dressing of downy sown cotton seeds, a modernized dressing machine UOSH-6, SP-3M, I-JS-8/L brand ("Yubus" Spain) is used (Fig. 1.13), or a set of equipment KPH-6 [89].

Fig. 1.13. I-JS-8/L downy seed dressing (Eubus, Spain)

In the seed duster I-JS-8/L the seed dosing unit is batch-operated, the necessary amount of slurry is applied to a certain portion of seeds, which is set by the operating staff. If the seed throughput is changed, the amount of slurry has to be adjusted again. Therefore, the quality of seed dressing depends on the qualification of the worker.

Complexes KPH-6, KPS-15 and KPS-19 have complete sets of systems for preparation of working solutions of dressing agents and their dosed supply for dressing [90]. For preparation of working suspension (solution) of dressing agent and its metered supply for dressing, the scheme shown in Figure 1.14 is used.

The system works as follows. In the tank 3 (Fig.1.14) the required amount of water is poured and at constant stirring the calculated amount of dressing agent is loaded.

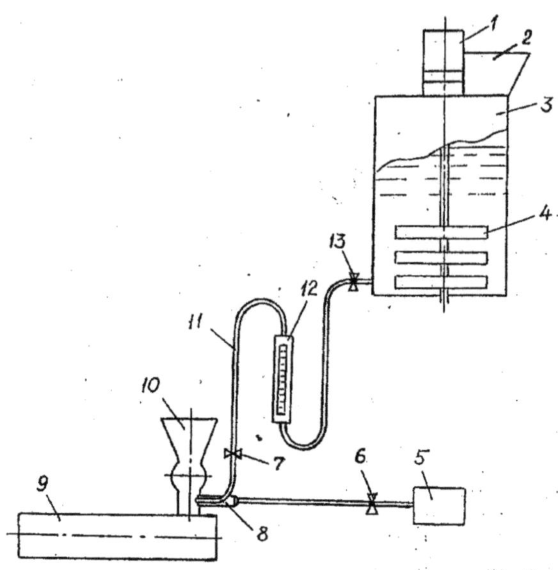

Fig.1.14. Scheme of simplified system of preparation of working solutions of dressing agents and their metered supply for dressing:1-motor-reducer; 2-pipe of preparation supply; 3-capacity (tank) of preparation and flow of suspension; 4-irrigator; 5-compressor; 6-vent; 7-vent; 8-nozzle; 9-etching machine; 10-seed doser with a hopper; 11-pipeline supply of dressing agent suspension; 12-rotameter; 13-crane.

Stirring is continued until the dressing agent is dissolved or a homogeneous suspension is obtained.

Slurry is supplied to dressing agent 9 by spraying with nozzle 8 after opening valve 13 and regulating its flow rate with valve 7. Slurry flow rate is controlled by means of rotameter

The disadvantage of this system is the difficulty in maintaining the slurry level above the dressing, which requires more careful control of the slurry flow rate, i.e. the position of the rotameter float 12.

In dressing shops of the Republic for dressing of downy cotton seeds the seed dosing hoppers [91] are introduced in which as a dosing mechanism the scheme developed on the basis of patents [92, 93] is applied (Fig.1.15).

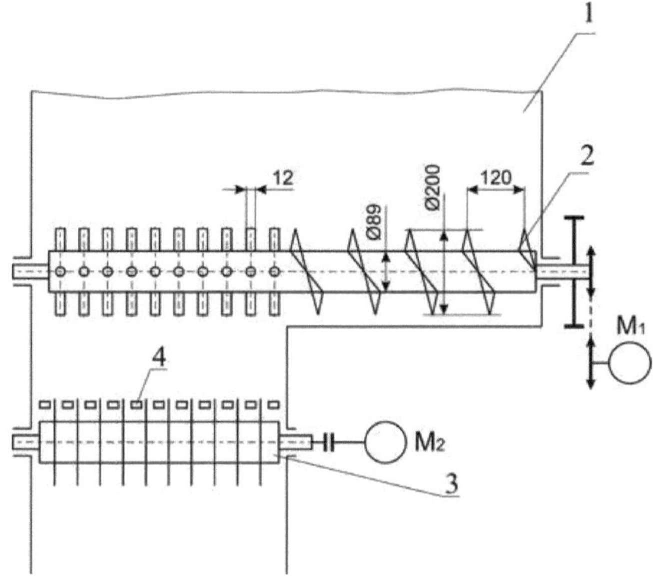

Fig.1.15 Schematic diagram of the dosing mechanism of the hopper-dosing unit

1- hopper; 2- combination rollers; 3- saw cylinder; 4- spike.

The disadvantage of this metering unit is the vaulting on top of the saw cylinder due to the mismatch between the seed feeding by the stake drums and the capacity of the saw cylinder. In addition, after the hopper of the metering unit the seeds are lifted to the top with the help of the bucket elevator of seeds, so the seeds are fed to the dressing chamber unevenly.

The author in the laboratory of JSC "Paxtasanoat ilmiy markazi" in order to eliminate the above-mentioned disadvantages of etchants has developed an improved scheme of the etching unit and produced an experimental sample [94, 95, 96, 97, 98, Annexes 1 and 2] (Fig. 1.16, 1.17, 1.18 and 1.19):

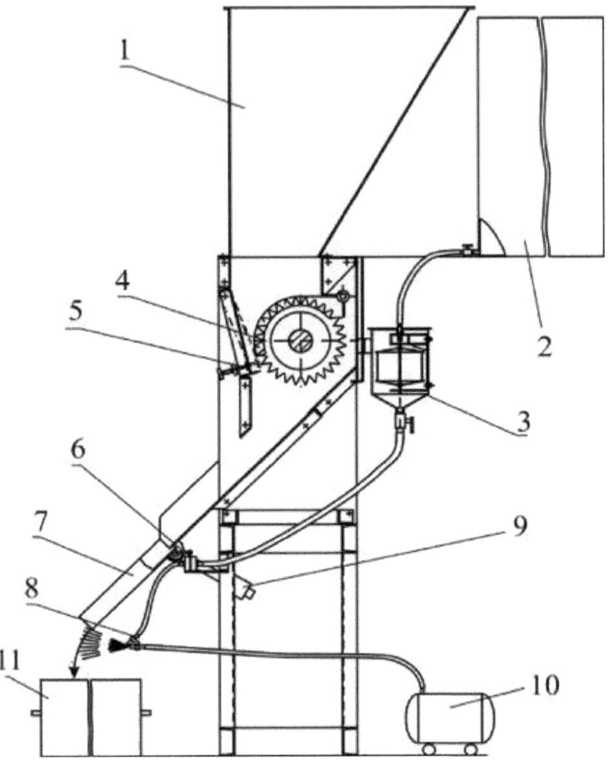

Fig.1.16 Principal scheme of the developed seed dressing machine of downed sowing seeds: 1-hopper; 2-tank for suspension; 3-pressure tank; 4-dispenser of sowing seeds; 5-flap; 6-crane; 7- swinging tray; 8-nozzle; 9-counterweight; 10-compressor; 11-drum for seed mixing.

Fig.1.17. General view of the experimental dressing unit of downy sown cotton seeds

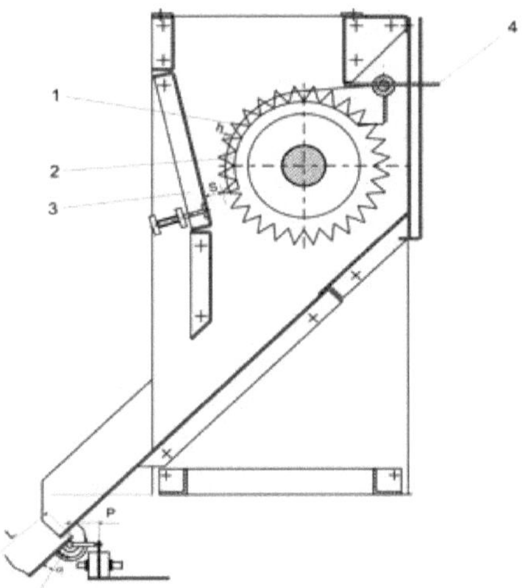

Fig.1.18. Schematic diagram of improved and simplified metering unit of downed seeds: 1-star cylinder; 2-column; 3-adjusting wall; 4-adjusting rod.

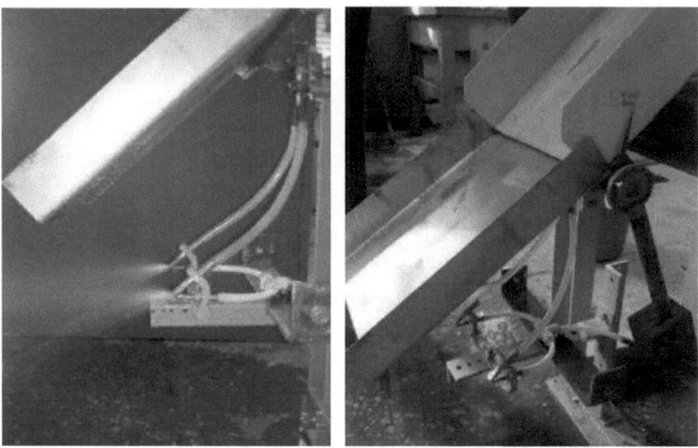

Fig.1.19. General view of the automatic stabilizer of the amount of dressing agent suspension with a nozzle depending on the seed output (right) and formation of suspension droplets with a nozzle (left).

As a result of preliminary studies, the following indicators of dropped seed dressing agent were obtained:

Table 1.7

Seed capacity, kg/hour	4000
Power requirement, kW	0,75
Seed metering cylinder rotation speed, rad./min	60
Seed metering cylinder diameter, mm	250
Overall dimensions, mm: -length -width -height	 642 370 2100
Weight, kg	250

Numerically, the quality of dressing is determined by three indicators: completeness of dressing, uniformity of distribution of preparations on individual seeds, the degree of retention (sometimes the inverse value is used - the degree of shattering).

The first indicator depends on subjective factors, in particular the personnel working on the dressing, and must therefore be constantly monitored by specialists. The completeness of the dressing is determined [3]:

$$\Pi = \frac{Q_{ж}}{H} \times 100\% \qquad (1.1)$$

where $Q_{ж}$ - is the actual weight of the preparation on seeds, kg/t;

H - recommended rate of consumption, kg/t.

The remaining two indicators are little influenced by subjective factors and generally hold up well.

1.6. Statement of the research problem

In connection with the above, the task of the present work is to develop and study the process of dressing downy seed cotton seeds, which allows to improve the quality of pre-sowing seed treatment with pesticides.

In order to accomplish the task at hand, it is necessary to:

1. Study of cotton seeds prepared for dressing, as an object of research.
2. Theoretical and experimental studies of the selected design of the dressing device and substantiation of the main parameters and modes of operation of the dressing device in order to ensure the required completeness and uniformity of dressing.
3. To make and experimentally investigate the selected design of dressing device in laboratory and production conditions, to determine its agrotechnical, energy, technical and economic evaluation and to implement it in production

1.7 Conclusions

1. Of the known methods of controlling cotton diseases, which are transmitted through seeds from year to year, the most effective at present and in the near future is chemical dressing of the seed surface from pathogens.

2. There are two main types of technical devices for treatment: flow type machine - the supply of preparation and seed to the treatment chamber is continuous; batch type machine - a certain measured amount of preparation is supplied to the treatment chamber for a certain weight of seed. It is considered that flow type machines have higher productivity.

3. The quality of dressing depends on many different factors. The analysis shows that the factors determining the quality of crop seed dressing can be grouped into four main groups, physical and mechanical properties of seeds, physical and chemical properties of the dressing agent, technological factors and factors depending on the design of the dressing agent.

4. Sprayers used in practice mostly disperse the working liquid with the preparation into drops of various sizes, i.e. they form polydisperse systems. In the best case it is possible to regulate the average particle size, and large weight fractions of different-sized particles reduce the effectiveness of treatment and adversely affect the uniformity and area of coverage of the seed surface with the dressing agent.

5. Many types of technologies and technical means for seed dressing of agricultural crops have been developed and introduced into production. However, until now there are no perfect solutions to improve the completeness of dressing, i.e., the compatibility of the amount of dressing agent supplied with the productivity of dressed seeds.

II. THEORETICAL BASIS FOR CALCULATING THE PARAMETERS OF THE PLANT FOR COTTON SEED DRESSING

2.1. Justification of parameters of the seed catching and dragging zone in the recommended feeding scheme

The recommended seed dresser provides for regulation of seed feeding capacity by changing the values of the drum tine protrusion from the grates of the plant. Fig.2.1 shows the schemes of location (orientation) of cotton seeds in the space between the tines at changing the height of the tines protrusion from the grate. Fig. 2.1 *a* shows the scheme of location of the seed at h<7.0 mm. Taking into account that the size of the seed reaches (10,0÷12,0) mm and its orientation between the tine space, two variants can be considered. In the first variant the center of gravity of the seed is outside the outer circle of the upper teeth of the drum. In this case, the seed will not be captured and dragged by the toothed drum. In the second embodiment, the center of gravity of the seed will be in the inter-tooth space of the drum. In this case the seed will be dragged into the feeding zone. Therefore, at h<7.0 up to 50% of the seed will be caught and dragged into the feed zone. Fig. 2.1 b shows the scheme at which h=7,0 mm and seeds will be dragged necessarily in one layer. Increasing the height of teeth protrusion from the grate correspondingly increases the amount of seeds dragged into the feeding zone.

When increasing h up to 10.0 mm (see Fig. 2.1 c), the amount of seeds dragged into the feeding zone increases 1.5 times. Increasing h to 13.0 mm allows cotton seeds to be pulled through in two or more layers.

It is important to study the process of capturing and dragging the seed by the drum teeth. Fig. 2.1 e shows the calculation scheme, in which the acting forces on the seed in static position are presented [91].

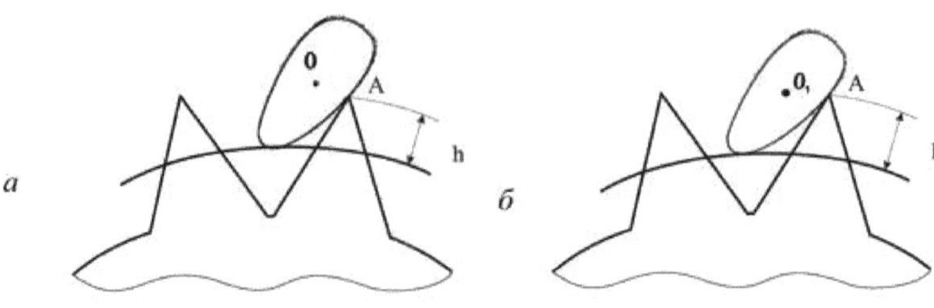

a- external height of teeth from the grate h<7.0 mm;

b - at h=7.0mm;

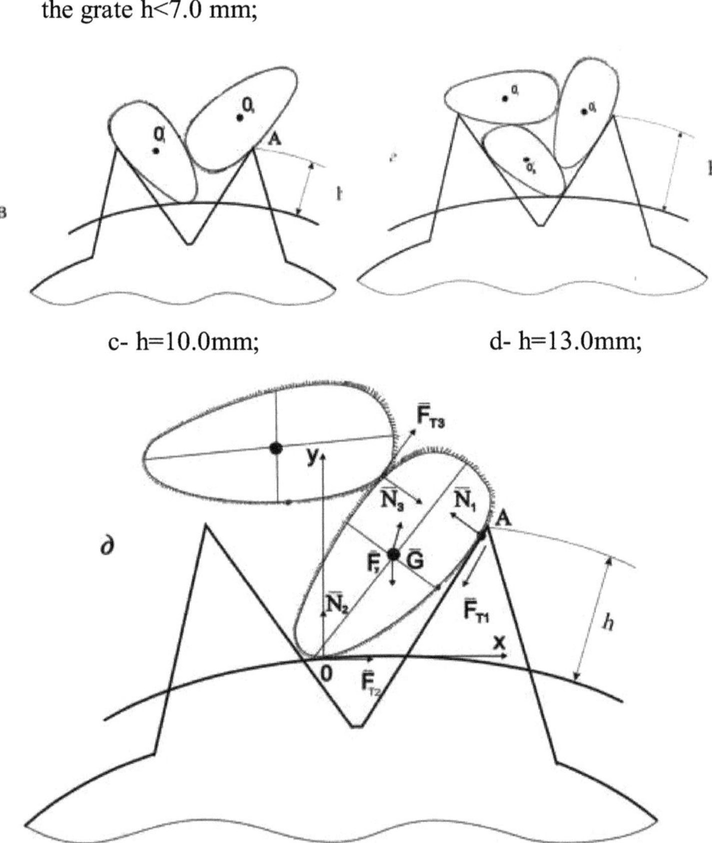

c- h=10.0mm;

d- h=13.0mm;

Fig.2.1 Seed arrangement schemes between tines of different values of grate tine protrusion: d-calculation scheme of seed interaction with drum tines.

Theoretically considered the condition of equilibrium of the seed in the inter-tooth space. Let's determine the condition of tipping of the seed when it is caught and carried by the drum teeth. From the calculation scheme (Fig. 2.(e)) follows that the following forces act on it: $\overline{G}$-weight force, $\overline{N}, \overline{F_{T_1}}$-reaction and friction forces between the seed and the surface of the drum tooth; $\overline{N_2}, \overline{F_{T_2}}$-reaction force and friction force between the seed and the surface of the grate; $\overline{N_3}, \overline{F_{T_3}}$- reaction and friction forces between the seed and neighboring seeds in the zone of capture and dragging into the feeding zone. In addition, inertia and centrifugal force acts on the captured seed.

Given that the seed does not move on the tooth surface, the inertia force will be zero. The centrifugal force will be:

$$F_y = m\omega^2 R \tag{2.1}$$

where m-seed mass; ω- angular velocity of the toothed drum; *R*-radius of location of the center of mass of the seed relative to the axis of rotation of the drum.

Considering the condition of equilibrium of the seed, taking the moments from the forces acting on the seed relative to point A we have:

$$Gh_G + N_3h_3 - N_2h_2 - F_{ц}h_{ц} - F_{T_1}h_1' + F_{T_2}h_2' = 0 \tag{2.2}$$

where, $h_G, h_{ц}, h_2, h_3, h_1', h_2'$-shoulders of the corresponding forces relative to the point A of the seed resistance and the surface of the drum tooth.

In order for the seed to remain in the interdental space, i.e. for the seed to be trapped and pulled through, it is necessary to fulfill the condition:

$$Gh_G + N_3h_3 - F_{T_2}h_2' \geq N_2h_2 + F_{Ц}h_{Ц} + F_{T_3}h_3' \quad (2.3)$$

Uchitivaya:

$$G = mg;\ F_{T_2} = f_2N_2;\ F_{T_3} = f_3N_3;\ h_2' = h$$

Determine the value of the protrusion of the drum tooth from the grate:

$$h \geq \left[\frac{h_2}{f} + \frac{N_3}{fN_2}\left(f_3h_3' - h_3\right) + \frac{m\left(\omega^2Rh_{Ц} + gh_G\right)}{fN_2}\right] \quad (2.4)$$

At numerical reaction (2.4) the following values of parameters were taken into account: f=0,3; f_3 =0,4; m=(0,12÷0,18)-10^{-3} kg, ω=6,28c^{-1} ; R=0,125m; g=9,8 m/s^2 ; N_2 =(0,4÷0,7)H; N_3 =(0,3÷0,5)H.

Analysis of the solution of problem (2.4) shows that with increasing the number of seeds interacting in the zone of their capture and dragging the values of N_2 and N_3 can increase by 2.0÷3.0 times. When dragging one layer of seeds N_2 and N_3 are up to 0,7 N and 0,5 N respectively. At dragging two-layer flow of seeds N_2 and N_3 reach 1,25 N and 1,0 N, and at (2,5÷3,0) layer flow of cotton seeds N_2 and N_3 reach 2,0 N and 1,5 N respectively. Thus according to (2.4) the height of the drum teeth protrusion from the grate at single layer flow of seed feeding is (7,0÷7,5)×10^{-3} m, at double layer flow of seed feeding the height of the teeth protrusion is (11,0÷13,0)×10^{-3} m, and at (2,5÷3,0) layer flow of seed dragging by the drum teeth the value of h should be more than (15,0÷17,0)×10^{-3} m. Taking into account experimental studies at productivity (4,0÷4,5) t/h recommended values of h are (7,0÷10,0)×10^{-3} m.

2.2 Determination of the trajectory of the seed to the inclined ruler of the dressing machine

In the recommended design the seeds are fed to the inclined guide by a toothed cylinder. At the same time, taking into account the initial flight velocity of seeds, they fall on different parts of the guide. It is important to determine the trajectory of seeds falling to the inclined plane of the guide from the change of system parameters.

The calculation scheme for determining the seed motion is shown in Fig. 2.2. In this case, the weight force and air resistance force mainly act on the seed. To determine the largest trajectory of the seed movement, we do not neglect the air resistance forces (due to their smallness).

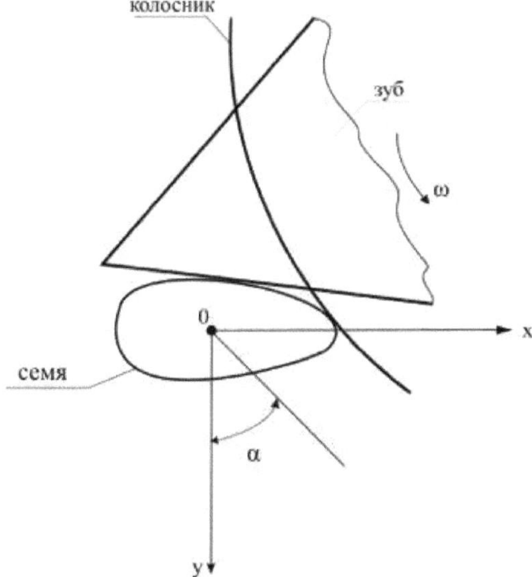

Fig.2.2 Calculation scheme

Then the system of differential equations describing the motion of the seed has the form [99]:

$$m\frac{d^2x}{dt^2} = 0$$

$$m\frac{d^2y}{dt^2} = mg \quad (2.5)$$

where, m-mass of seed; $t -$ time; x, y —coordinates of the seed's motion; g —acceleration of free fall.

In the initial moment:

$$x_0 = 0; \frac{dx}{dt} = V_0 cos\alpha;$$

$$y_0 = 0; \frac{dy}{dt} = V_0 sin\alpha \quad (2.6)$$

In doing so.

$$V_0 = \omega(R - a) \quad (2.7)$$

where, ω -is the angular velocity of the toothed drum; R —is the outer radius of the toothed drum; a is the distance from the outer circumference of the drum to the center of mass of the seed.

By twice integrating the first equation (2.5) we obtain:

$$\frac{dx}{dt} = c_1; \; x = c_1 t + c_2 \quad (2.8)$$

Considering the initial conditions at t=0, we determine the integration constants:

$$c_2 = V_0 cos\alpha; \; c_1 = 0$$

Then we obtain the following solution of the problem:

$$x = V_0 t cos\alpha; \; \dot{x} = \frac{dx}{dt} = V_0 cos\alpha \quad (2.9)$$

Integrating twice the second equation (2.5) we obtain:

$$V_y = \dot{y} = gt + c_1^{'} \; ;$$

$$y = \frac{gt^2}{2} + c_1^{'} t + c_2 \quad (2.10)$$

Taking into account the initial conditions according to (2.6), we determine the integration constants:

$$t = 0;\ c_1' = 0;\ c_2' = V_0 sin\alpha$$

(2.11)

Then the solution of the second equation (2.5) will be:

$$\dot{y} = V_y = \frac{dy}{dt} = gt + V_0 sin\alpha;$$

$$y = \frac{gt^2}{2} + V_0 t sin\alpha \qquad (2.12)$$

Taking into account that the height of seed fall from the toothed drum to the inclined guide is known (within certain limits), from the second equation (2.12) it is possible to determine the time of seed flight:

$$t = \frac{V_0 sin\alpha \pm \sqrt{V_0^2 sin^2\alpha + 2gh}}{g} \qquad (2.13)$$

The obtained value of t substituting into the solution (2.9) we have:

$$x = V_0 \left[\frac{V_0 sin\alpha \pm \sqrt{V_0^2 sin^2\alpha + 2gh}}{g}\right] cos\alpha \qquad (2.14)$$

When numerically solving the problem to determine the trajectory of the seed movement, the distance from the axis of the drum to the inclined tray of the structure 0.230 m, and in the area of the flap 0.264 m was taken into account. Given that the frequency of rotation of the toothed drum 60 rpm, we take α within

$\left(\frac{\pi}{6} \div \frac{\pi}{4}\right)$.

The initial linear velocity of the seed will be:

$$V_0 = \omega(R - a) = (0{,}734 \div 0{,}740) \text{ м/с}$$

In this case, R=0.125 m; a=$(5,0 \div 6{,}5)\times10^{-3}$ m.

Numerical solution of the problem is presented in the form of graphs, trajectories of seed movement in the zone between the toothed drum and the inclined tray, which are shown in Fig.2.3. Analysis of the obtained trajectories of seed movement shows that with increasing the rotational speed of the toothed drum (see Fig.2.3a) increases the horizontal component of the trajectory. So, in the first dependence at angular speed of toothed drum 5,6 s^{-1} trajectory of seed movement will be non-linear and reaches the inclined tray at y=0,214 m and x=0,072 m, and at angular speed 7,0 s^{-1} coordinates of seed on the surface of the inclined tray reach to y=0,102 m and x=0,196 m. This means that with increasing angular speed of the toothed drum increases the coordinate of the trajectory of the seed on the horizontal axis, and decreases on the vertical axis. It should be noted that the increase in the coordinate of the seed trajectory on the *x-axis* leads to an increase in the values of movement of the seed on the tray, which is undesirable.

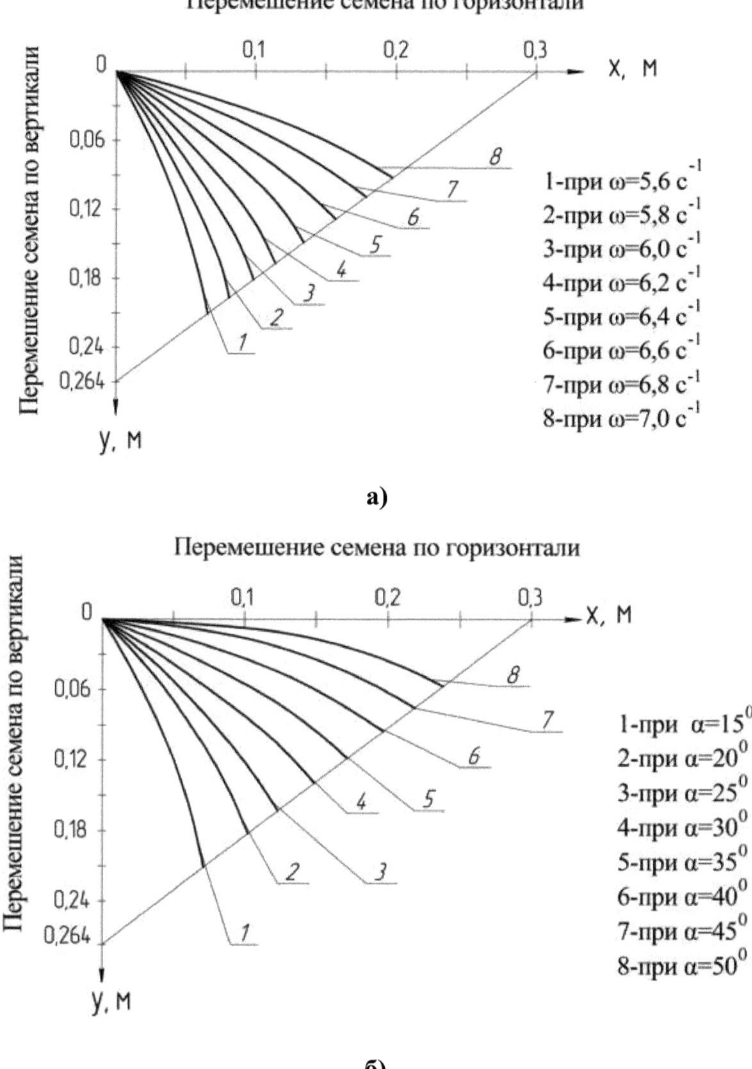

Fig. 2.3. Trajectory of cotton seed movement in the area between the toothed drum and the inclined tray: a - changing the speed of the toothed drum; b - changing the angle of seed ejection from the tines of the drum.

Increasing the coordinate on the vertical axis of seed movement can lead to significant accumulation of seeds in the zone of entrance to the dosing system, which is also not desirable. Therefore, the angular speed of the toothed drum (6,0÷6,35) with^{-1} , at which a more uniform distribution of seed on the inclined tray of the unit is provided, is considered the most acceptable.

Analysis of the graphs in Fig. 2.3 b shows that the increase in the angle of seed ejection from the drum teeth, depending on the phase of the tooth location leads to an increase in the X coordinate. At the same time, the greater the angle α of seed ejection, the greater the horizontal component of the vector of linear velocity of the seed at the initial moment of ejection. In order to ensure uniform distribution of seed on the inclined tray it is considered reasonable $\alpha=20^0 \div 30$.^{0}

2.3 Study of cotton seed movement on the inclined tray of the dressing machine

In the recommended design of the seed dressing machine it is important to ensure uniform feeding of seeds into the zone of suspension application. Therefore, it is reasonable to study the movement of cotton seeds on the inclined tray. Fig. 2.4 shows the design scheme of the inclined tray of the seed dresser. Seeds on the tray are mainly moved by the force of their weight. Part of the seeds can be moved by enveloping (rolling friction).

The following forces mainly act on the seed located on the surface of the inclined tray: weight force; friction force $\bar{G}$; the force of friction $\bar{F}_{тр}$ of the seed against the surface of the tray; $\bar{N}$-force of reaction; $\bar{F}_{и}$-force of inertia. If we designate coordinate axes X and Y, the seed will move only along the X axis, and along the Y axis=0.

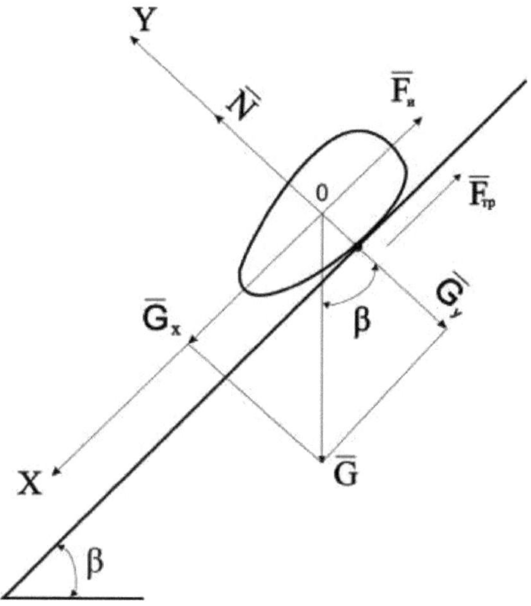

Fig. 2.4. Calculation scheme

In this case, the equations of motion of the seed along the axes according to [100] will be:

$$m_c\ddot{x} = Gsin\beta - fGcos\beta$$

$$m_c\ddot{y} = N - Gcos\beta$$

(2.15)

where, $m_c -$ seed mass; $(0,124\div0,13)\times10^{-3}$ kg; free fall acceleration g=9,81 m/s^2 ; tray inclination angle $\beta=45^0$; seed *friction* coefficient on the surface tray, *f=0,3*. Considering y=0, N = Gcosβ.

B initial moment $x_0 = 0, y_0 = 0$,

$$\dot{x} = \dot{x_0}cos\beta;\ \dot{y} = \dot{y_0}sin\beta$$

The first equation of the system of differential equations (2.15) will be:

$$m_c\ddot{x} = G(sin\beta - fcos\beta)$$

(2.16)

Integrating the differential equation (2.16) twice, we obtain:

$$\dot{x} = gt(sin\beta - fcos\beta) + C_1,$$

$$x = \frac{gt^2}{2}(sin\beta - fcos\beta) + C_1 t + C_2 \qquad (2.17)$$

Given the initial conditions at t=0; C_1 =0; C_2 =0, we have:

$$\dot{x} = gt(sin\beta - fcos\beta); x = \frac{gt^2}{2}(sin\beta - fcos\beta) \qquad (2.18)$$

According to the obtained dependencies in Fig. 2.3, the coordinate of the seed falling on the surface of the tray on the x-axis, as well as the length from the axis of the cylinder to the zone of suspension of the inclined tray $0.5\cdot10^{-3}$ m we determine the distance of movement of the seed on the inclined tray, (0.67÷0.7) m. This is the maximum distance of movement of the seed being in the rightmost position on the surface of the tray. On average, this distance will be within (0.40÷0.52) m.

From the second equation of the system (2.18) we can determine the time of seed movement along the tray:

$$t = \sqrt{\frac{2x}{m_c g(sin\beta - fcos\beta)}} \qquad (2.19)$$

Then substituting (2.19) into the first equation of the system (2.18) we obtain an expression for determining the seed velocity at the end of the inclined tray of the cotton seed dresser:

$$\dot{x} = \sqrt{2xg(sin\beta - fcos\beta)} \qquad (2.20)$$

Numerical solution of (2.20) is made at the following initial values of parameters: x = (0,40÷0,52) m; $\beta=30^0 \div 55^0$; g=9,81 m/s^2 , *f=0,2÷0,4*.

Fig. 2.5 a shows graphical dependences of change of seed velocity in the end of the inclined tray of the seed dresser on the change of the angle of inclination of the tray at variation of the friction coefficient between the seed and the surface of the tray. From the analysis of graphs it follows that with increasing the angle of inclination of the tray linear velocity of the seed at the end of the unloaded increases according to a nonlinear pattern. So, when the angle β increases from 30^0 to 55^0 at *f=0.4* the seed velocity increases from 1.21 m/s to 2.33 m/s, and at *f=0.2* the velocity V_c increases from 1.74 m/s to 2.55 m/s. Therefore, to increase the supply of seed to the suspension zone, it is necessary to increase the angle of inclination of the tray or decrease the friction coefficient between the seed and the surface of the tray. The recommended values of parameters are $\beta=40^0 \div 45^0$, *f=0,30* at which the productivity of seed dresser is provided within (4,0÷4,5) tons.

Fig.2.5 b shows graphical dependences of the change of seed velocity in the unloading zone from the increase in the coordinate x at different values of f. From the graphs it can be seen that with increasing the coordinate x from 0,40 m to 0,52 m at *f=0,4* the speed of the seed at the end of the tray to reach 2,33 m/s, and at f=0,2 the speed increases to 2,55 m/s. To ensure the productivity of cotton seed dressing within (4,0÷4,5) tons recommended value of the coordinate x are less than (0,48÷0,52) m at a gear drum speed of 60 rpm.

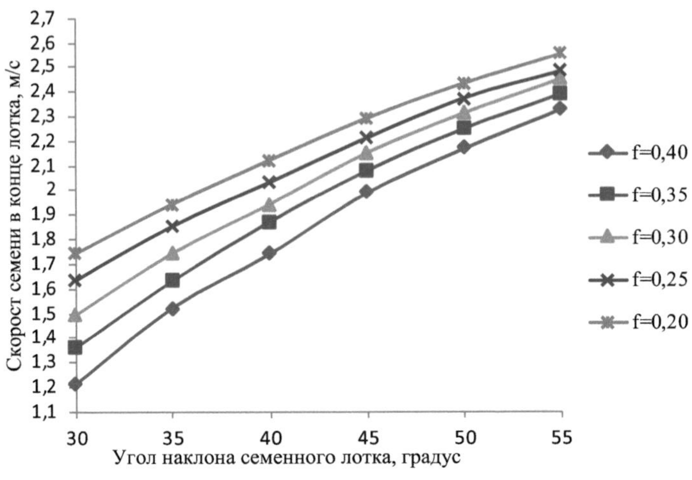

a)

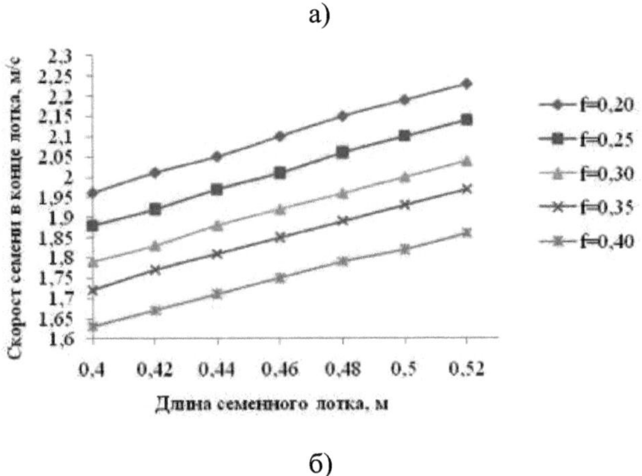

б)

Fig. 2.5. Graphical dependences of the change in the seed velocity at the end of the tray on the change in the angle of inclination of the tray (a) and on the value of seed displacement.

2.4 Analyzing the vibrations of the dressing tray

In the zone under consideration, up to 8333 pcs of seeds pass through the oscillating tray simultaneously. Taking into account that the mass of downed seeds is within (0,124÷0,132) gr. At the same time in the oscillating tray there will be seeds with total mass of (1,0÷1,08) kg of seeds depending on their pubescence.

To make a mathematical model of the oscillating system, Fig. 2.6 shows the calculation scheme.

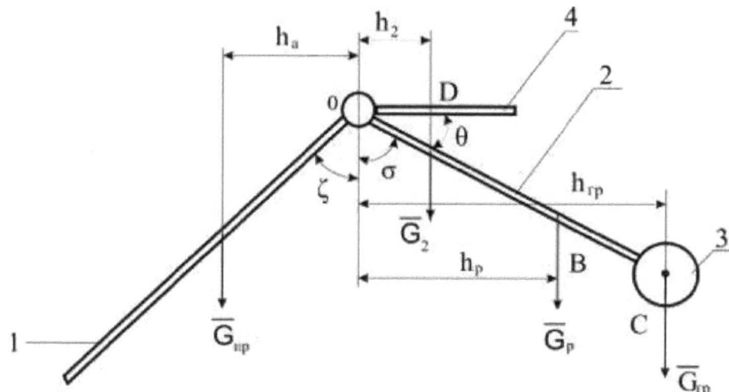

Fig. 2.6. Calculation scheme

1- tray; 2- load lever; 3- load (counterweight); 4- scale lever

If the mass of the tray itself is 1.3 kg, then the moment created by the forces of the weight of the seeds in the tray, taking into account the force of the weight of the tray will be:

$$M_1 = M_л + (M_c \pm \delta M_c)$$

(2.21)

Where, $M_л = \frac{1}{2} m_л g l_л sin\zeta$; $M_c = \frac{1}{2} n m_c g l_л sin\zeta$

$$\delta M_c = (0{,}10 \div 0{,}15) M_c sin\omega t$$

In this case, let us rewrite (2.21) in the following form:

$$M_1 = \frac{1}{2} g l_л sin\zeta_л (m_л + n m_c) + (0{,}05 \div 0{,}075) n m_c g l_л sin\zeta_л sin\omega t \qquad (2.22)$$

where, $l_л$ –is the length of the tray; ζ –tray rotation angle; m_c –seed mass; n –number of seeds simultaneously in the tray;

$m_л$ –tray mass; g – acceleration of free fall; γ –frequency of change of the number of seeds simultaneously in the oscillating tray.

The moment relative to the joint created by the counterweight with the adjusting weight according to the scheme will be:

$$M_2 = M_p + M_{гр} \quad (2.23)$$

где, $M_p = \frac{g}{2} l_p m_p \sin(\sigma + \zeta)$; $M_{гр} = \frac{g}{2} m_{гр} l_{гр} \sin(\sigma + \zeta)$

with that being said: $$M_2 = \frac{g}{2}\left(l_p m_p + l_{гр} m_{гр}\right) \sin(\sigma + \zeta) \quad (2.24)$$

where, l_p-is the length of the counterweight arm; m_p-mass of the counterweight arm; $m_{гр}$-mass of the load; $l_{гр}$-length of the arm of the lever part of the load location; σ-angle of inclination of the counterweight lever.

The torque generated by the lever connected to the gate valve of the dressing valve:

$$M_3 = \frac{g}{2} l_2 m_2 \sin(\sigma + \theta + \zeta) - \mathrm{M_k} \quad (2.25)$$

l_2-Length of arm with brackets; m_2-weight of the arm with brackets; -angle of inclination of the lever with brackets; $\mathrm{M_k}$-resistance to the lever pin on the side of the valve gate valve.

At each instant of time the tray, counterweight, lever arm with the slide relative to the axis of the installation is in equilibrium therefore [101]:

$$\sum_{i=1}^{n} M = M_1 - M_2 - M_3 - M_u = 0 \quad (2.26)$$

Taking into account the reduced moment of inertia of the three-lever arm, the calculations were performed according to the methodology given in [102], J_{np} =0.634 kg·m² . Then, taking into account the moment of inertia of the three-lever lever using the Lagrange equation of the II kind, we obtain the following differential equation describing the oscillations of the lever:

$$J_n\ddot{\zeta}_л = \frac{1}{2}gl_л sin\zeta_л(m_л + nm_c) + (0,05 \div 0,075)nm_c gl_л sin\zeta_л sin\omega t - \frac{g}{2}(l_p m_p + l_p m_{гp})\sin(\sigma + \zeta) - \frac{1}{2}l_2 m_2 \sin(\sigma + \theta + \zeta_n) - M_K \qquad (2.27)$$

It should be noted that the three-floating lever of the seed dresser oscillates with a small amplitude, we take $sin\zeta = \zeta$, then (2.27) is rewritten in the form:

$$J_n\ddot{\zeta}_л = \frac{1}{2}gl_л\zeta_л(m_л + nm_c) + (0,05 \div 0,075)nm_c gl_л\zeta sin\omega t - \frac{g}{2}(l_p m_p + l_p m_{гp})(\sin\sigma + \zeta cos\sigma) - \frac{1}{2}l_2 m_2[\sin(\sigma + \theta) + \zeta\cos(\sigma + \theta)] - M_K \qquad (2.28)$$

The numerical solution of (2.28) was performed on a PC at the following parameter values:

$J_{np} = 0,634$ $кг \cdot м^2$, $g = 9,81$ $^{м}/_{c^2}$, $l_л = 0,3$ $м$; $m_л = 1,3$ $кг$; $m_p = 0,12$ $кг$; $m_{гp} = 1,1$ $кг$; $m_k = 0,014$ $кг$; $l_p = 0,21$ $м$; $l_2 = 0,048$ $м$; $\sigma = 30^0 \div 45^0$; $\theta = 20^0 \div 25^0$; $M_K = 0,42$ $Нм$; $n = 8333$ *шт*; m_c=*(0,124÷0,132)·10⁻³ kg*.

Analysis of the obtained equation (2.28) shows that the three-arm lever of the seed dresser actually makes parametric oscillations during operation. In this case, the perturbing force from the fed seeds depends on the value of change of the oscillation node of the tray with seeds. On the basis of numerical solution of the problem were obtained regularities of oscillation of the tray of cotton seed dresser. Fig. 2.7 shows the regularities of the tray oscillations at different values of the system parameters. Fig.

2.7 (a), (b), (c) shows that with the increase of the mass of the load on the lever the amplitude and frequency of the tray oscillations decrease. Thus, at $m_{гр} = 1{,}3$ кг, $\sum m_c = 1{,}08$ кг the range of oscillations of the tray reaches 3.1^0 and accordingly the period of oscillations decreases to 0.71 s.

Important is the influence of changing the production rate of the dressing agent on the tray oscillation (see Fig. 2.7 d, e). Fig. 2.8 shows graphical dependences of the change of the tray oscillation range on the total mass of seeds in the dressing tray. The analysis of the obtained graphical dependences shows that with the rise of sunrise $\sum m_c$ the tray oscillation range increases according to the non-linear regularity. Thus, when the total mass of seeds increases from 0.4 kg to 2.4 kg, the tray oscillation range increases from 1.21^0 to 3.09^0 when the mass of the load is 1.5 kg. When the mass of the load is reduced to 1.0 kg, the oscillation range of the seed dressing tray increases from 1.29^0 to 5.84^0 . At the same time, it should be noted that the oscillation of the tray with a larger amplitude can lead to a decrease in the reliability of operation, as well as the seeds falling out of the tray.

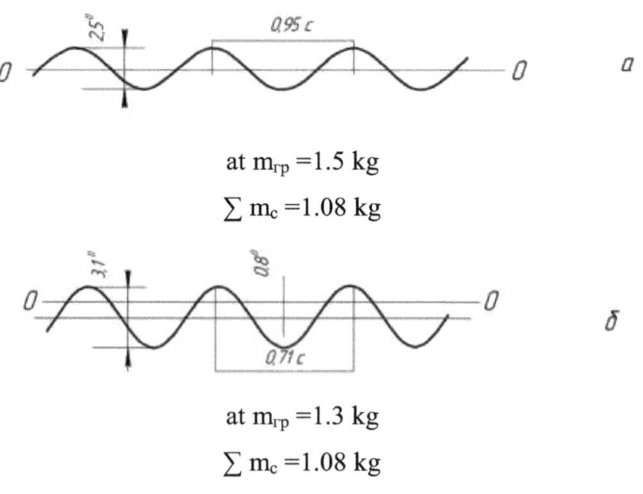

at $m_{гр}$ =1.5 kg
$\sum m_c$ =1.08 kg

at $m_{гр}$ =1.3 kg
$\sum m_c$ =1.08 kg

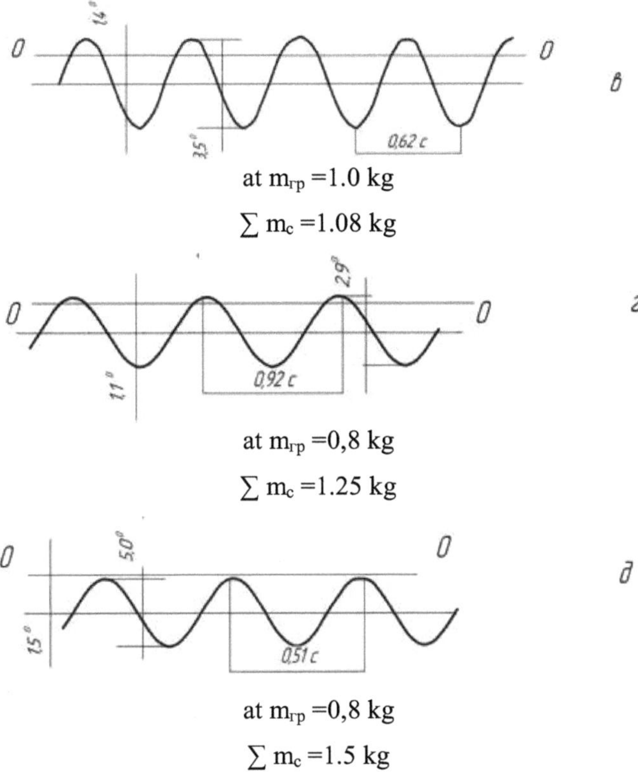

Fig. 2.7. Regularities of oscillations of the cotton seed dressing trough

Tray oscillations with a spread of less than 2.0^0 do not allow effective seed movement in the tray. Therefore, the recommended values are: $m_{гр} = (1,0 \div 1,3)$ кг; $\sum m_c = (1,1 \div 1,5)$ кг at $n_p = (4,0 \div 4,5)$ т/ч.

Fig. 2.9 shows graphical regularities of change of oscillation range of seed dressing tray from change of lever angle with load at variation of load weight values. From the analysis of the plotted graphs it is seen that the increase of angle σ leads to increase $\Delta\zeta$ by nonlinear regularity. Thus, at the mass of the load of 1.6 kg the increase of the angle σ from 25^0 to 50^0 leads to an increase in the range of oscillations of the tray with seeds from 0.51^0 to 2.03^0 , and with a decrease in the mass of the load $\Delta\zeta$ varies from 1.23 to 5.67^0 . This is explained by the fact that the increase of angle σ leads to an increase in the impact shoulder of the load relative to the hinge of the tray. This also increases the influence of the uniform component of the disturbing force. Recommended values of the angle of the lever arm with the load in relation to the seed tray are (75^0 ÷80).^{0}

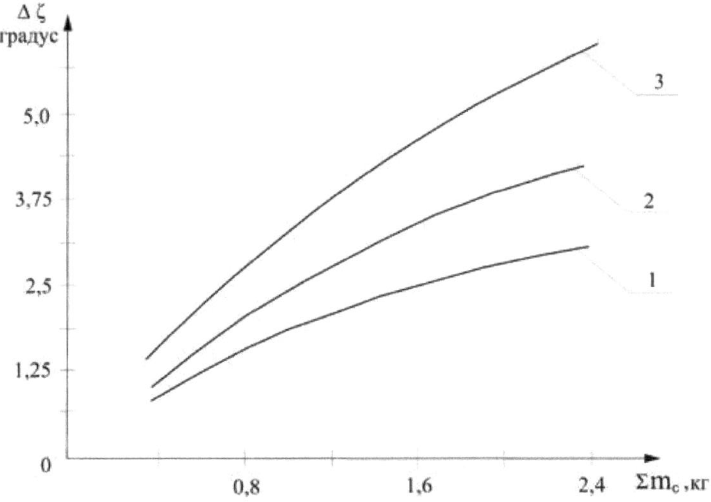

Fig. 2.8. Dependences of change of angular oscillation range of the dressing tray on the change of seed weight in the tray

1-Pri $m_{гр}$ =1.5 kg; 2-Pri $m_{гр}$ =1.3 kg; 3-Pri $m_{гр}$ =1.0 kg.

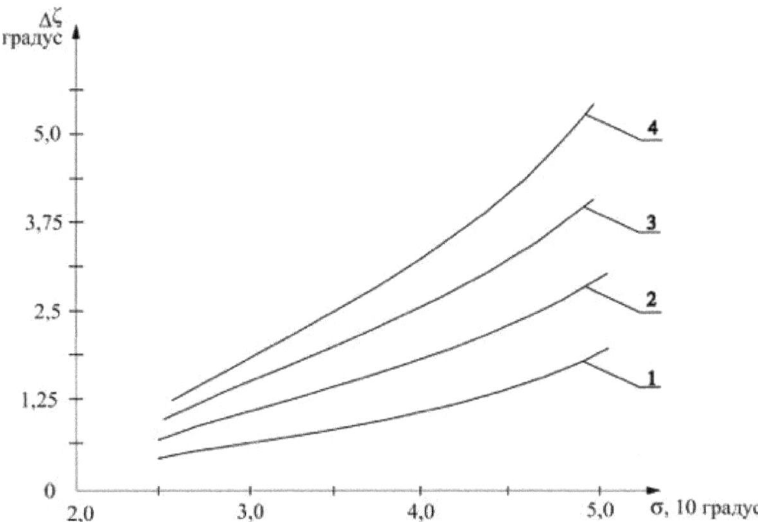

Fig. 2.9. Graphical dependences of the change of the oscillation range of the seed dressing tray on the change of the angle of the lever with a load 1-at $m_{гр}$ =1.6 kg; 2-at $m_{гр}$ =1.3 kg; 3-at $m_{гр}$ =1.0 kg; 4-at $m_{гр}$ =0.7 kg.

2.5 Conclusions

1. The formula for calculating the distance of the drum tooth protrusion from the grate is obtained. To ensure the productivity of seed dressing within (4,0÷4,5) tons it is recommended the value of tooth protrusion from the grate within (7,0÷10,0)-10^{-3} m.

2. The equations of motion of the seed when falling on the inclined guide are obtained. Formulas for determining the trajectory of seed movement are obtained by analytical method. Formulas for determining the time of seed movement are obtained.

3. The regularities of changes in the trajectory of cotton seeds in the zone between the toothed drum and the inclined guide from changes in the speed of rotation of the toothed drum and the angle of seed ejection in the initial zone are constructed.

4. It is revealed that increasing the coordinate of the seed trajectory along the x-axis leads to an increase in the values of seed movement along

the tray, which is undesirable. Increasing the coordinate on the vertical axis of the seed movement can lead to a significant accumulation of seeds in the zone of entrance to the dosing system. The most acceptable is considered angular speed of toothed drum (6,0÷6,35) with^{-1} , at which more uniform distribution of seed on the inclined tray of the installation is provided.

5. It is established that at the angle of seed ejection in the initial zone leads to an increase in the horizontal component of the initial seed velocity vector. It is recommended $\alpha = 20^0 \div 30^0$.

6. The equations describing the movement of the seed on the inclined tray of the dresser are obtained. The formula for determining the time of seed movement along the inclined tray of the dressing machine is derived.

7. Graphical dependences of the change of seed velocity in the tray on the change of inclination angle and on the value of seed displacement are plotted. To increase the seed feeding to the suspension zone it is necessary to increase the tilt angle of the tray or decrease the friction coefficient between the seed and the tray surface. Recommended values of parameters are: $\beta=40^0 \div 45^0$, $f=0,30$, at which the productivity of seed dressing machine is provided within (4,0÷4,5) tons.

8. To ensure the productivity of cotton seed dressing within (4,0÷4,5) tons, the recommended coordinate values are x≤(0,48÷0,52) m at toothed drum rotation speed of 60 rpm.

9. A mathematical model describing small oscillations of the tray with seeds of the seed dresser is obtained. By numerical solution of the problem the regularities of the seed dressing tray oscillations change are obtained. It is revealed that the amplitude and frequency of the tray oscillations decrease with the increase of the weight mass on the lever. At

$m_{гр} = 1{,}3$ *кг*, $\sum m_c = 1{,}08$ *кг* the range of oscillations of the tray reaches 3.1^0 and, accordingly, the oscillations decreases to 0.71 s.

10. The graphical dependences of the change of the tray oscillation range from the total seed mass of the seed mass in the dressing tray are constructed. Analysis of the obtained graphical dependencies shows that with increasing $\sum m_c$ the tray oscillation range increases according to non-linear regularity. When increasing the total mass of seeds from 0.4 kg to 2.4 kg, the range of oscillation of the tray increases from 1.21^0 to 3.09^0 at a mass of 1.5 kg. The recommended values are: $\mathrm{m_{гр}} = (1{,}0 \div 1{,}3)$кг; $\sum \mathrm{m_c} = (1{,}1 \div 1{,}5)$ кг at $\mathrm{n_p} = (4{,}0 \div 4{,}5)$т/ч.

11. Graphical regularities of change of the oscillation range of the seed dressing tray from the change of the lever angle with the load at variation of the load weight values are constructed. At the weight of the load 1,6 kg the increase of angle σ from 25^0 to 50^0 leads to increase of the oscillation range of the tray with seeds from $0{,}51^0$ to $2{,}03^0$, and with decrease of the mass of the load $\Delta\zeta$ varies from 1.23 to 5.67^0 . Recommended values of the angle of the lever with the load in relation to the seed tray are (75^0 ÷80).^{0}

III. METHODOLOGY AND RESULTS OF EXPERIMENTAL RESEARCH

3.1 General program of experimental studies

The overall program of experimental research involved conducting:

1) lab experiments on:

(a) Determination of the throughput capacity of the pipette depending on its main parameters;

b) investigation of the slurry stabilizer providing supply of the required amount of working fluid to the nozzle regardless of the hydraulic head of the slurry flow tank;

c) investigation of the suspension flow rate depending on the angle of inclination of the rocking tray of downed seeds;

d) investigation of suspension flow rate depending on the value of the distance from the axis of the swinging tray to the place of fixing the automatic suspension valve in the horizontal surface.

2) laboratory and production experiments on:

(a) Determination of the seed treatment capacity of the seed dresser;

b) investigation of the completeness of seed dressing;

c) study of uniformity of seed dressing.

All laboratory, laboratory and production studies were carried out with prepared for sowing (calibrated and cleaned from dust) pubescent sowing seeds of cotton breeding variety C-6524, reproduction R2, residual pubescence 7.8 %, moisture 8.6 %, mechanical damage 3.8 % and working solution of suspension "P-4" prepared according to the recommendations on cotton seed dressing [3].

3.2 Methodology of laboratory experiments

3.2.1 Methodology for determination of downed seed metering capacity

The dressing performance depends on the throughput capacity of the downed seed metering unit.

The purpose of laboratory and production experiments is to determine the throughput capacity of the downed seed metering unit. To conduct experiments to determine the throughput capacity of the metering unit, a laboratory stand of the downy seed metering unit was made (Figure 3.1). Throughput capacity of the downed seed metering unit was determined depending on the following factors:

-value of exit of teeth of toothed cylinder 1 from the comb h (Fig.3.1) from 7 to 16 mm, change interval 3mm;

- distance between teeth of toothed cylinder 1 and adjustable wall 3- S from 10 to 30 mm, change interval 5 mm;

- gear cylinder rotation speed n, taken as 60 rpm.

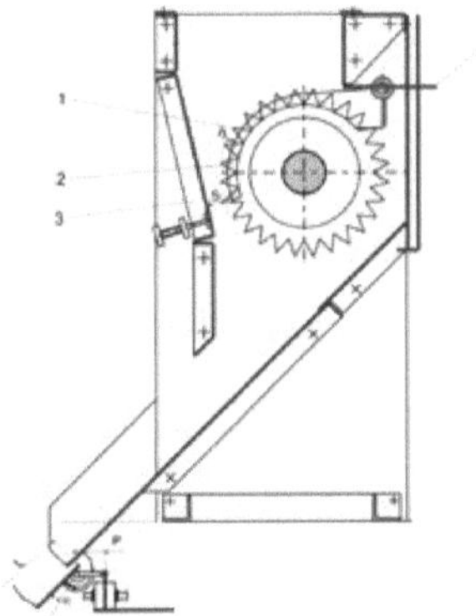

Fig.3.1 Schematic diagram of downed seed metering unit

1-toothed cylinder; 2-scallop; 3-adjustable wall; 4-pull rod for adjustment.

The experiments were carried out in the following sequence. The hopper is half filled with seeds, starts the gear motor of the metering unit, set the speed of the gear cylinder at 60 rpm, set the required values of the above factors, in steady mode of operation for 1 minute in a bag collected seeds passing through the metering unit. Seeds are weighed on the scales HD 4050-300 with an error margin of ±0.1 kg. Then the hopper capacity is calculated according to the following formula:

$$Q_6 = 60q' \qquad (3.1)$$

where Q_6 - is the throughput capacity of the metering unit, kg/h;

q' - amount of seeds poured out through the metering unit during the experiment period, kg.

In this order, the experiments are repeated in triplicate and the average value of the throughput capacity of the metering unit at the selected variant of the experiment is calculated.

Then the hopper is half filled again and the experiments are repeated in this sequence for the following variants of the metering unit operation.

This technique can be used to determine the throughput capacity of metering units of any geometric shape, with any design and with seeds of different cotton seed selections.

3.2.2 Methodology for measuring the performance of the slurry stabilizer

It is known that according to the law of hydraulics of liquids, when the liquid, for example, in a cylindrical container, the hydrostatic pressure to the walls in the horizontal direction does not change and remains constant. However, the hydrostatic pressure to the bottom of the container, the pressure in the vertical direction changes, the magnitude of this pressure depends on the amount of liquid in this container.

Hydrostatic pressure depends on the height of the liquid h, the magnitude of the acceleration of free fall g, the area of the bottom of the vessel s and the specific density of the liquid ρ:

$$P = \frac{F}{S} = \frac{\rho g S h}{S} = \rho g h \tag{3.2}$$

According to this formula, the higher the height of the liquid, the higher the hydrostatic pressure at the bottom of the tank. The flow of the dressing agent suspension is carried out by an elastic pipe with the help of a valve, which is usually installed closer to the bottom of the tank. The required amount of dressing agent suspension is set with this valve. If the operator has been adjusting the valve at the same volume of slurry in the tank, then over time (as the amount of slurry in the tank decreases or as the hydrostatic pressure from the working slurry to the bottom of the tank decreases), the amount of slurry delivered through the valve will decrease. If the operator adjusted the valve with less slurry in the tank, the opposite phenomenon occurs.

In order to stabilize the hydrostatic pressure to the bottom of the tank, the author has developed a scheme and manufactured an experimental sample of etchant (Fig. 3.2) with a stabilizer of hydrostatic pressure to the bottom of the flow tank according to the scheme shown in Fig. 3.3.

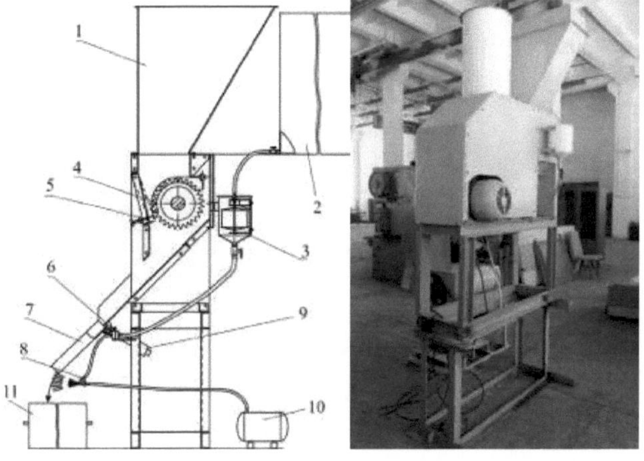

Fig. 3.2 Schematic diagram and general view of the experimental dressing plant with a stabilizer of hydrostatic pressure of liquid

1- hopper; 2- tank for slurry; 3-pressure tank; 4- seed seed dispenser; 5- flap; 6-crane; 7- oscillating tray; 8- nozzle; 9-counterbalance; 10- compressor; 11-drum for mixing seeds.

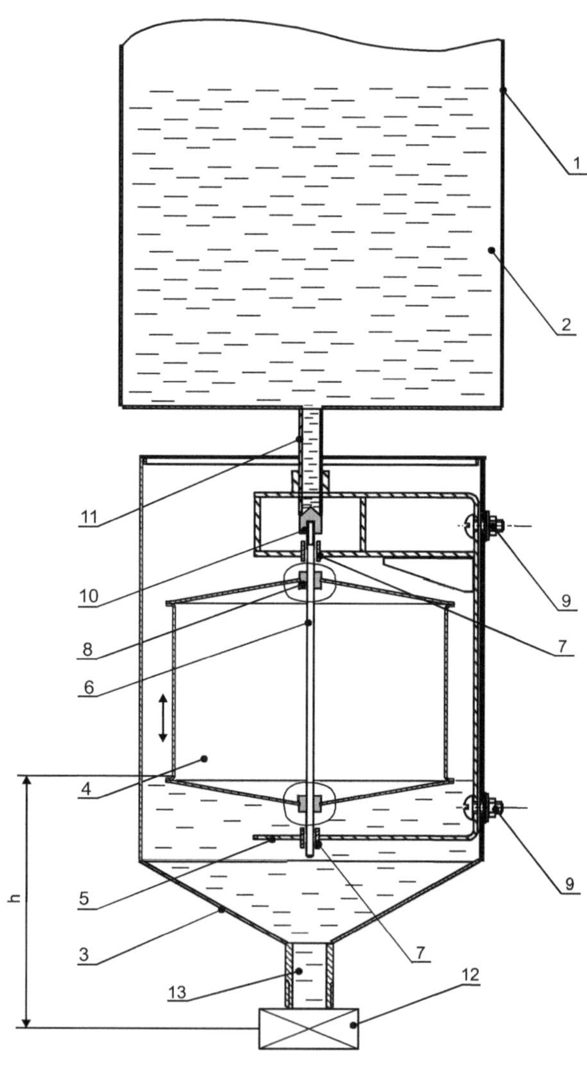

Fig. 3.3. Schematic diagram of the hydrostatic pressure stabilizer together with the working suspension tank of the dressing agent

The hydrostatic pressure stabilizer 3 (Fig. 3.3) is installed downstream from the flow tank 1, inside which the float 4 is installed on the bracket 5 with the help of the seal 8 and screw 9 on the vertical axis 6. The position of the float 4 on the axis 6 is adjusted vertically with the help of the seal 8 and screw 9. At the upper end of the axis 6 is installed automatic valve 10 opening or closing orifices on the pipe 11 communicating with the flow tank 1. From the stabilizer the working suspension of the dressing agent flows to the metering unit 12 (conditionally) through the pipe 13 installed in the lower part of the hydrostatic pressure stabilizer.

The hydrostatic pressure stabilizer works as follows:

From the flow tank 1 the dressing agent suspension 2 flows through the pipe 11 into the hydrostatic pressure stabilizer 3. As the working suspension accumulates, the float 4 together with the vertical axis 6 rises to the top. The automatic valve 10 located on the axis 6 also rises and closes the opening in the pipe 11. If the dressing will not work (no slurry flow through the pipe 13) the position of float 4 does not change and the hole in the pipe 11 will be closed by automatic valve 10. When the dressing agent is working, the slurry valve 6 will be opened (Fig. 3.2) and there is a flow of working slurry from the stabilizer 3. At the same time automatic valve 10 together with float 4 goes down and opens opening in pipe 11. The working slurry starts to flow from the flow tank 1 into the stabilizer 3. As the working slurry accumulates, the float 4 rises, for example, to a height h, and the automatic valve 10 closes the flow of working slurry from the container 1. Thus, inside the stabilizer all the time the slurry height is at the same set level (height), so the influence of hydrostatic pressure of the working slurry of the dressing agent on the flow rate set after the stabilizer of the valve 6 (Fig.3.2) is prevented.

To check the operation of the hydraulic pressure stabilizer, experiments were carried out according to the following methodology. For the flow capacity of the working suspension of the dressing agent a cylindrical tank with a capacity of 500 liters was selected, the inner diameter of the pipe for the suspension outlet is 15 mm (Fig. 3.4.).

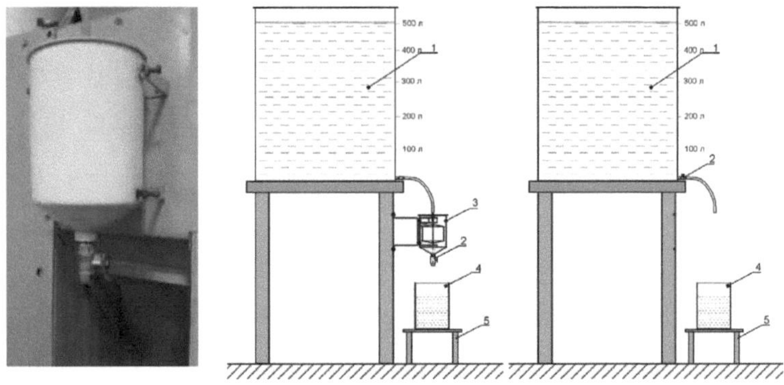

1-Tank with a capacity of 500 l; 2-valve; 3- hydrostatic pressure stabilizer; 4- measuring tank; 5- metal stand.

Fig.3.4 Schematic diagram to the method of determining the hydrostatic pressure stabilizer operation.

Experiments were carried out in two variants, the first variant without installation of hydrostatic pressure stabilizer, the second variant with installation of hydrostatic pressure stabilizer after the flow tank.

In the first variant after the flow tank on the pipe for the outlet of suspension is installed valve and the required flow rate of suspension was set at the amount of suspension on the flow tank equal to 100 liters. Within 10 seconds at the established mode of suspension flowing through the valve was collected in a measuring cup and determined the actual flow rate of suspension. Repetition of each variant of the experiment was three times. Then the same experiments were repeated at the amounts of suspension on the flow capacity 150, 200, 250, 300, 350, 400, 450, 500 liters.

In the second variant, a hydrostatic pressure stabilizer was installed downstream of the flow tank on a pipe with an inner diameter of 13 mm. After the stabilizer installed a valve for the flow of slurry and the required flow rate of slurry was set at the amount of slurry on the flow tank equal to 100 liters. During 10 seconds at the established mode the suspension flowing through the valve was collected in a measuring container and the actual flow rate of suspension was determined. Then the same experiments as in the first variant were repeated with the amount of suspension on the flow container 150, 200, 250, 300, 350, 400, 450, 500 liters.

3.2.3 Methodology for determining the suspension flow rate depending on the angle of inclination of the swinging tray of downed seeds and depending on the value of the distance from the axis of the swinging tray to the place of fixing the automatic suspension valve in the horizontal surface

It is known that according to recommendations [3], for example, when dressing downy seeds with preparation P-4, the consumption of suspension for 1 ton of seeds should be within 25-30 liters. In the experiments exactly this preparation was used and the average consumption was taken as 27,5 liters. The parameters of the shaking tray were chosen experimentally-length 300 mm, width 260 mm and sidewall height 50 mm.

During the descent of the seed, the rocker trough changes its angle depending on the capacity of the seed metering unit, i.e. on the mass of seed simultaneously on the surface of the rocker trough.

Swinging batten 1 is interconnected with automatic suspension valve 2 installed at some adjustable distance in the horizontal surface (Fig.3.5).

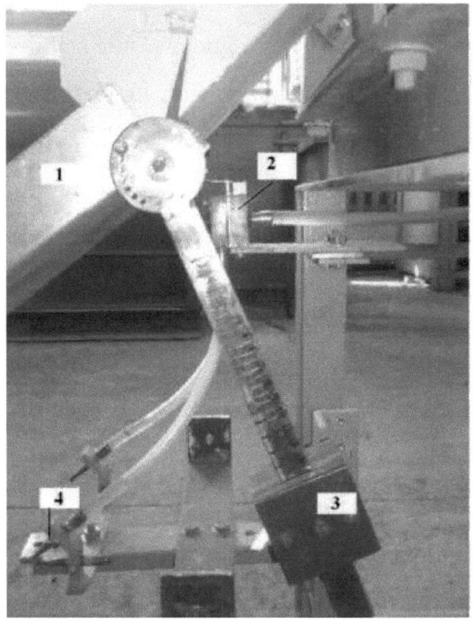

1-seed tray; 2-automatic valve; 3-counterweight; 4-nozzle

Fig.3.5 General view of automatic control of slurry flow rate

Counterweight 3 (Fig.3.5) serves for repayment of frequency of oscillation of oscillating tray 1 and smooth change of its angle from the mass of seeds simultaneously located on the surface of the seed tray.

During the experiments the following values from the axis of the swinging tray to the automatic tap of the dressing suspension in the horizontal surface were checked: L=28 mm; 32 mm; 36 mm; 40 mm; 44 mm; 48 mm; 52 mm;

Seed metering capacity was measured between 2500-4000 kg/hour at intervals of 500 kg/hour.

Initial state of the automatic valve regardless of the distance value L in closed position when there is no seed on the surface of the swing tray.

Seeds of selection variety C-6524 with 8.5% pubescence were used for experiments. Repetition of each variant of experiments is threefold.

3.3.2 Methodology of research of dressing completeness and uniformity of working fluid distribution

The most important indicators that characterize the quality of work of the dressing machine are the completeness of dressing and uniformity of distribution of the working liquid over the mass of seeds and over the surface of each individual seed.

Five bags of cotton seeds prepared for sowing, weighing 50 kg each, were prepared for conducting experiments to study the completeness of dressing. The working suspension prepared according to the instructions was poured into the tank of the dressing agent. From each batch of seeds prepared for dressing 10 portions of seeds from different places are taken into bins of approximately the same size and their weight is determined.

Experiments are carried out in the working mode of dressing at a capacity of 4 t/h in the following sequence. 50 kg of seeds are poured into the hopper and the treater is started. The flow rate of the working suspension is set to the rate of 27.5 l/t. Dusted seed doser is set to a capacity of 4 t / h and conducted experiments timing the time until the emptying of the hopper.

Then, 10 portions of seeds of approximately the same size are taken from batches of treated seeds from different locations and their weight is determined. Then the arithmetic mean of each batch of seeds before and after dressing is calculated. The difference in the masses of treated ($m_{\text{пп}}$) and untreated ($m_{\text{нп}}$) seeds we take as the actual consumption ($q_{жс}$) of working suspension. This consumption is recalculated per 1 ton of seeds according to the formula:

$$Q_{жс} = 1000\frac{q_{жс}}{m_{нп}} \qquad (3.3)$$

where $q_{жс}$- is actual flow rate of working fluid in the tested portion of seeds, g;

$m_{нп}$ - seed weight before dressing in the tested seed portion, g.

In calculations we assume that the density of the working slurry is equal to the density of water, respectively 1 liter of working slurry is equal to the mass of 1 kg.

Then the fullness of dressing is calculated according to the formula

The completeness of the dressing is determined by :

$$П = \frac{Q_{жс}}{H} \times 100\% \qquad (3.4)$$

where $Q_{жс}$ - actual weight of the preparation on seeds, kg/t;

H - recommended maximum consumption rate, kg/t.

It is considered satisfactory if the coefficient of variation C_v does not exceed 30%.

In the same sequence experiments are carried out at the norms of working slurry consumption 25, 27.5 and 30 l/t and productivity 4 t/h.

In parallel with the determination of the completeness of dressing, the uniformity of distribution of the working suspension of the dressing agent over the seed mass and the surface of individual seeds was determined.

To determine the uniformity of distribution of the working suspension by seed weight, the results of weighing the weight of seeds after dressing in each portion were used.

The uniformity of drug distribution to seeds is estimated by the formula:

C

$$= \frac{S'}{Q_ж} \times 100\% \qquad (3.5)$$

where S′ - is the mean square deviation;

$Q_ж$ - arithmetic mean of the quantities determined in individual batches (kg/t)

For this purpose, statistical processing of data was carried out. Arithmetic mean, mean square deviation in separate portions and coefficient of variation were determined by formula (3.8). The less will be the coefficient of variation, the higher will be the uniformity of distribution of working suspension by seed mass.

3.4 Methods of processing experimental results studies

Experimental data of laboratory and laboratory-production studies were processed by known methods of mathematical statistics [103].

One of the tasks of statistical processing of experimental data is to find some values that characterize the sample statistical population. Sufficient information about the experiment can be obtained by the following characteristics: mean value $-\bar{x}$; standard deviation (mean square deviation) - S; standard error (error of the mean) - S; standard error (error of the mean) - V. $S_{\bar{x}}$; coefficient of variation -V.

The most widely used characteristic is the arithmetic mean, which is the quotient of dividing the sum of the values of all variants by their number:

$$\bar{x} = \frac{x_1 + x_2 + x_3 + \cdots + x_n}{n} = \frac{\sum x}{n} \qquad (3.6)$$

One of the most important statistical characteristics is the mean square deviation, which characterizes the dispersion of variant values with respect to the mean of the distribution, i.e. the arithmetic mean:

$$S = \sqrt{\frac{\sum(x-\bar{x})^2}{n-1}} \qquad (3.7)$$

where$x$ is the value of individual variants;

$\bar{x}$ - arithmetic mean;

n is the number of variants.

The standard deviation is a named number and is expressed in the same units as the measurement data. This makes it difficult to compare different-sized traits to assess the degree of their variation. The relative indicator of variability of the material under study can be calculated in the form of the coefficient of variation:

$$V = \frac{S}{\bar{x}} \cdot 100\% \qquad (3.8)$$

The obtained values of experimental data are displayed in the form of graphs plotted on a computer in the operating environment "MS Windows XP" using the program "Microsoft Excel 2007".

3.5 Results of laboratory experiments performed

3.5.1 Determination of seed hopper metering capacity

The purpose of the laboratory experiments is to determine the throughput capacity of the seed hopper. As mentioned earlier, the throughput capacity of the seed hopper must be greater than the maximum capacity of the seed dressing machine. According to the results of theoretical studies, a maximum throughput of 4 t/h is required to ensure the maximum throughput. Laboratory experiments to determine the throughput capacity of the seed hopper dosing unit were carried out according to the method described in subsection 3.2.1.

The following objectives are set for the experimental experiments:

- without interrupting the dispenser;
- accurate dosing of the dropped seeds;
- uniform feeding of the dropped seeds from the device;
- improve the accuracy of correlation of the seed metering unit with the amount of slurry.

Preliminary tests have shown that 3 factors influence the performance of the pipette:

- saw blade protrusion of the saw cylinder from the grate.
- distance "S" from the saw teeth of the saw cylinder to the adjusting wall;
- saw cylinder speed, n.

Experiments were then conducted in the following type of parameters:

- saw blade protrusion of the saw cylinder from the grate;
- distance "S" from the saw teeth of the saw cylinder to the adjusting wall;
- saw cylinder rotation speed n, 60 rpm was taken as unchangeable.

In the experiments, pubescent sowing seeds (untreated) of the breeding variety C-6524, 2-reproduction, with mechanical damage of 3.8 %, pubescence of 7.8 % and moisture content of 8.6 % were used [104].

The results of the experiments are shown in Tables 3.1-3.5 and Figure 3.6.

Table 3.1.

Influence of the saw cylinder saw teeth protrusion from the grate on the performance of the seed metering unit at the distance "S" from the saw cylinder saw teeth to the regulating wall equal to S=10mm;

№		Capacity, kg/hour			

	Protrusion of saw cylinder saws from the grate, mm	Repetition of experiments			Average value of throughput capacity, kg/h	Mean square deviation, S	Coefficient of variation, V
		1	2	3			
1	7	2300	2420	2380	2367	61,1	2,6
2	10	3376	3295	3370	3347	45,1	1,3
3	13	4310	4392	4320	4342	44,8	1,0
4	16	5284	5361	5333	5326	39,0	0,7

Table 3.2.

Influence of the saw cylinder saw teeth protrusion from the grate on the performance of the seed doser at the distance "S" from the saw cylinder saw teeth to the regulating wall equal to S=15 mm;

№	Protrusion of saw cylinder saws from the grate, mm	Capacity, kg/hour			Average value of throughput capacity, kg/h	Mean square deviation, S	Coefficient of variation, V
		Repetition of experiments					
		1	2	3			
1	7	3250	3360	3300	3272	67,1	2,0
2	10	4160	4190	4200	4184	20,8	0,5

3	13	4970	5010	5000	4994	20,8	0,4
4	16	5950	5890	5930	5924	30,6	0,5

Table 3.3.

Influence of the saw cylinder saw teeth protrusion from the grate on the performance of the seed metering machine at the distance "S" from the saw cylinder saw teeth to the regulating wall equal to S=20 mm;

№	Protrusion of saw cylinder saws from the grate, mm	Capacity, kg/hour Repetition of experiments			Average value of throughput capacity, kg/h	Mean square deviation, S	Coefficient of variation, V
		1	2	3			
1	7	4210	4190	4195	4198	10,4	0,2
2	10	4950	4980	5050	4994	51,3	1,0
3	13	5670	5500	5580	5584	85,0	1,5
4	16	6510	6530	6490	6510	20,0	0,3

Table 3.4.

Influence of the saw cylinder saw teeth protrusion from the grate on the performance of the seed metering machine at the distance "S" from the saw cylinder saw teeth to the regulating wall equal to S=25 mm;

№		Capacity, kg/hour			

	Protrusion of saw cylinder saws from the grate, mm	Repetition of experiments			Average value of throughput capacity, kg/h	Mean square deviation, S	Coefficient of variation, V
		1	2	3			
1	7	5010	4995	5060	5022	34,0	0,7
2	10	5650	5580	5600	5610	36,0	0,6
3	13	6240	6190	6200	6210	26,4	0,4
4	16	6800	6840	6790	6810	26,4	0,4

Table 3.5.

Influence of the saw cylinder saw teeth protrusion from the grate on the performance of the seed metering machine at the distance "S" from the saw cylinder saw teeth to the regulating wall equal to S=30 mm;

№	Protrusion of saw cylinder saws from the grate, mm	Capacity, kg/hour Repetition of experiments			Average value of throughput capacity, kg/h	Mean square deflection, S	Coefficient of variation, V
		1	2	3			
1	7	5560	5620	5600	5594	30,6	0,5
2	10	6050	5980	6000	6010	36,0	0,6
3	13	6570	6500	6490	6520	43,6	0,7
4	16	6980	7010	7100	7030	62,4	0,9

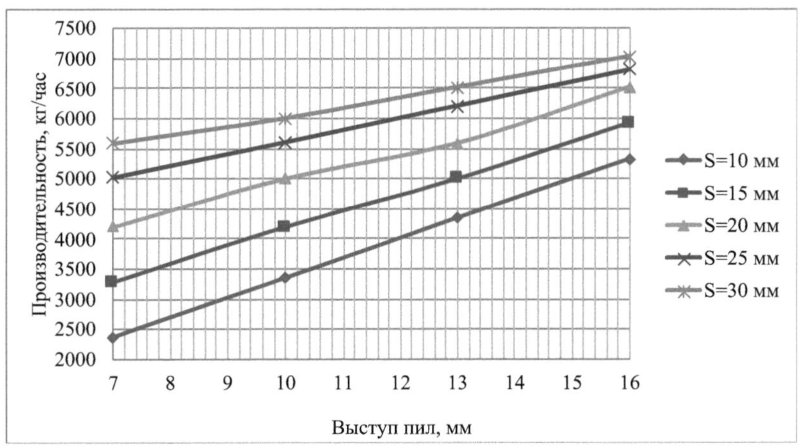

Fig.3.6 Influence of saws overhang above the grate on seed feeding performance

It can be seen from the experiments that changing the distance "S" from the saw teeth of the saw cylinder to the regulating wall and protrusion of the saw cylinder from the grate comb affect the performance of the metering unit. By increasing the distance "S" up to 30 mm the performance of the metering unit increases, but the seeds are not evenly discharged, i.e. in clusters. According to our requirement and at high output the lowered seeds must be fed to the dressing uniformly.

During the experiments, the analysis of different positions of the metering unit adjustment showed that the distance from the saw teeth of the saw cylinder to the regulating wall S = 20 mm and at the protrusion of the saws of the saw cylinder from the grate comb by 7-10 mm, the productivity of the metering unit is not less than 4000 kg/hour. At the same time seed feeding is uniform, which generally satisfies the conditions.

3.6 Measurement results of slurry stabilizer performance

As indicated in Section 3 of this work experiments on measuring the work of the working suspension stabilizer were carried out in two variants, the first variant without installation of the hydrostatic pressure stabilizer, the second variant with the installation of the hydrostatic pressure stabilizer after the flow tank [105].

In the first variant after the flow tank on the pipe for the outlet of the slurry a valve is installed and the necessary flow rate of the slurry was set at the amount of slurry on the flow tank equal to 100 liters. During 10 seconds at the set mode the suspension flowing through the valve was collected in a measuring cup and the actual flow rate of the suspension was determined. Repetition of each variant of the experiment was three times. Then the same experiments were repeated with the amount of suspension on the flow capacity 150, 200, 250, 300, 350, 400, 450, 500 liters.

In the second variant, a hydrostatic pressure stabilizer was installed on the pipe with inner diameter downstream of the flow tank. After the stabilizer installed a valve for the flow of slurry and the required flow rate of slurry was set at the amount of slurry on the flow tank equal to 100 liters. During 10 seconds at the set mode the suspension flowing through the valve was collected in a measuring container and the actual flow rate of suspension was determined. Then the same experiments as in the first variant were repeated at the amounts of suspension on the flow capacity 150, 200, 250, 300, 350, 400, 450, 500 liters. The results of the experiments are given in Tables 3.6 and 3.7.

Table 3.6.

Results on determination of slurry flow rate depending on the amount of slurry in the flow tank at operation without hydraulic pressure stabilizer

Quantity of slurry on the flow tank, l	Slurry flow rate for 10 sec., l				Mean square deviation, S	Coefficient of variation, V
	Repetition of experiments					
	1	2	3	Average value		
100	0,31	0,33	0,32	0,32	0,01	3,125
150	0,35	0,33	0,34	0,34	0,01	2,941
200	0,34	0,35	0,36	0,35	0,01	2,857
250	0,37	0,36	0,35	0,36	0,01	2,777
300	0,37	0,38	0,39	0,38	0,01	2,631
350	0,40	0,45	0,35	0,40	0,05	12,5
400	0,41	0,43	0,42	0,42	0,01	2,381
450	0,47	0,43	0,45	0,45	0,02	4,444
500	0,48	0,49	0,47	0,48	0,01	2,083

According to the results of Table 3.6 and 3.7 we can conclude that the developed hydraulic pressure stabilizer provides more uniform supply of working slurry to the dressing agent regardless of the amount of slurry in the flow tank. When working without the hydraulic pressure stabilizer the flow rate of slurry at increasing the amount of slurry in the flow tank (when setting the required flow rate at the amount of 100 liters on the flow tank) varies in a wide range and leads to excessive consumption of working slurry and poor quality dressing of cotton seeds.

Table 3.7.

Results on determination of slurry flow rate depending on the amount of slurry in the flow tank when working with hydraulic pressure stabilizer

Quantity of slurry on the flow tank, l	Slurry flow rate for 10 sec., l				Mean square deviation, S	Coefficient of variation, V
	Repetition of experiments					
	1	2	3	Average value		
100	0,310	0,330	0,320	0,320	0,010	3,125
150	0,320	0,310	0,330	0,320	0,010	3,125
200	0,328	0,323	0,324	0,325	0,0026	0,8
250	0,340	0,320	0,330	0,330	0,010	3,030
300	0,330	0,340	0,320	0,330	0,010	3,030
350	0,335	0,330	0,325	0,330	0,005	1,515
400	0,335	0,333	0,337	0,335	0,0014	0,418
450	0,338	0,341	0,341	0,340	0,0017	0,5
500	0,340	0,339	0,341	0,340	0,001	0,294

3.7 Determination of the slurry flow rate depending on the angle of inclination of the pumping trough and the distance from the axis of the pumping trough to the automatic valve in the horizontal surface

The following values of the studied factors were taken during the experiments:

1. The following dimensions of the swing tray are accepted; length-300 mm; width-260 mm; height of the side walls of the tray-50 mm.

2. Distance from the axis of the swinging tray to the place of fixing the automatic slurry supply valve in the horizontal surface, L=28 mm; 32 mm; 36 mm; 40 mm; 44 mm; 48 mm and 52 mm.

3. counterweight weight: 500 gr; 700 gr; 900 gr and 1100 gr.

The working suspension of the preparation P-4 was used in the tests.

The test work was primarily performed on the swing tray oscillation repayment.

As it is known when the seed passes through the rocking chute, oscillations are generated which changes the carbon α.

These fluctuations affect the degree of uneven opening of the slurry valve, which in turn affects the uneven supply of slurry for seed dressing. Therefore, in order to repay the oscillation of the tray we change the mass of counterweight, its place on the lever and the angle of inclination of the lever to the tray. Experimental studies determined the following parameters: the rational mass of counterweight-1100 grams, the distance to the place of its installation on the lever - 210 mm and the angle of inclination of the lever to the rocking tray - 75 . ^{0}

After determining the above parameters, the main research was carried out to determine the influence of the value of the distance from the axis of the swinging tray to the place of fixing the automatic slurry valve in the horizontal surface to supply the required amount of slurry at the doser capacity of 2500, 3000, 3500 and 4000 kg/hour.

The results of the studies are summarized in Tables 3.8, 3.9, 3.10, and 3.11.

Table 3.8.

Influence of the distance from the axis of the swing chute to the automatic valve on the slurry flow rate at metering capacity of 2500 kg/hour

№	Distance from the axis of the swing tray to the automatic slurry valve, mm	Repetition of experiments			Average value, l/hour	Mean square deviation, S	Coefficient of variation, V
		1	2	3			
1	28	43,8	44,0	43,3	43,7	0,36	0,82
2	32	50,3	49,8	50,5	50,2	0,36	0,72
3	36	55,9	56,0	56,7	56,2	0,43	0,76
4	40	61,9	62,7	62,9	62,5	0,53	0,85
5	44	68,1	69,1	68,9	68,7	0,53	0,77
6	48	75,4	74,7	74,9	75,0	0,36	0,48
7	52	81,6	81,3	80,7	81,2	0,46	0,57

The analysis of the obtained results showed that the most acceptable distance from the axis of the swinging tray to the place of fixing the automatic valve of suspension in the horizontal surface for supplying the required amount of suspension at the doser capacity of 4000 kg/hour can be taken as 44 mm to 48 mm. Because, the norm of suspension consumption according to "Recommendations on cotton seeds dressing" - for dressing of 1 ton of lowered sowing seeds is set in the amount of 25-30 liters. If the specified distance is reduced from 44 mm, the slurry supply will be less than the norm, and if its increase is more than 48 mm, the slurry supply will be more than the norm [106].

Table 3.9.

Influence of the distance from the axis of the swing chute to the automatic valve on the slurry consumption at metering capacity of 3000 kg/hour

№	Distance from the axis of the swing tray to the automatic slurry valve, mm	Repetition of experiments			Average value, l/hour	Mean square deviation, S	Coefficient of variation, V
		1	2	3			
1	28	52,1	52,7	53,0	52,6	0,46	0,87
2	32	59,7	59,9	60,4	60,0	0,36	0,60
3	36	67,6	67,2	67,7	67,5	0,26	0,38
4	40	75,0	74,8	75,5	75,1	0,36	0,48
5	44	82,8	81,9	82,8	82,5	0,52	0,63
6	48	90,1	89,5	89,8	89,8	0,30	0,33
7	52	97,0	97,9	97,6	97,5	0,46	0,47

Based on the results of the experiments, the following optimum parameters of the machine for dressing of the dropped seeds were determined: the protrusion of the saws of the saw cylinder from the grate h=7-10 mm, the distance from the teeth of the saws of the saw cylinder to the regulating wall S=20 mm, the rotational speed of the saw cylinder n=60 rpm, the length of the swinging tray 300 mm, the distance from the axis of the swinging tray to the automatic valve for the supply of suspension L=44-48 mm and the weight of the counterweight - 1100 g.

Table 3.10.

Influence of the distance from the axis of the swing chute to the automatic valve on the slurry consumption at metering capacity of 3500 kg/hour

№	Distance from the axis of the swing tray to the automatic slurry valve, mm	Repetition of experiments			Average value, l/hour	Mean square deviation, S	Coefficient of variation, V
		1	2	3			
1	28	61,0	61,4	61,2	61,2	0,20	0,33
2	32	70,3	70,2	70,4	70,3	0,10	0,14
3	36	78,4	78,8	78,9	78,7	0,26	0,33
4	40	87,7	87,2	87,6	87,5	0,26	0,30
5	44	96,0	96,5	96,1	96,2	0,26	0,27
6	48	105,5	105,0	105,1	105,2	0,26	0,25
7	52	114,0	113,5	113,6	113,7	0,26	0,23

Table 3.11.

Influence of the distance from the axis of the swing chute to the automatic valve on the slurry flow rate at doser capacity of 4000 kg/hour

№	Distance from the axis of the swing tray to the automatic slurry valve, mm	Repetition of experiments			Average value, l/hour	Mean square deviation, S	Coefficient of variation, V
		1	2	3			
1	28	70,1	70,5	70,6	70,4	0,26	0,37
2	32	80,2	80,0	79,8	80,0	0,20	0,25
3	36	89,8	90,4	90,1	90,1	0,30	0,33

4	40	99,7	100,1	100,2	100,0	0,26	0,26
5	44	110,0	110,5	110,4	110,3	0,26	0,23
6	48	120,0	119,8	119,6	119,8	0,20	0,17
7	52	124,9	124,7	124,2	124,6	0,36	0,29

3.8 Investigation of dressing completeness and uniformity fluid distribution

The main qualitative indicators of the dressing process are the completeness of dressing and uniformity of distribution of the working fluid over the seed mass and over the surface of each individual seed [54].

To study the qualitative parameters of dressed seeds of grain crops, methods of quantitative determination of active substances of preparations based on the technique of gas-liquid chromatography are used [14]. However, this method of investigating the quality of seed dressing requires expensive laboratory equipment, laboratory and qualified specialists. It should also be noted that there are no officially approved methods for determining the quality of cotton seed dressing. Therefore, for the study and determination of completeness and uniformity of dressing we have compiled the methods described in subsection 3.2.2.

As can be seen from the experimental data, the completeness of dressing at the studied (from 25 to 30 l/t) suspension flow rates on the experimental dressing machine is 85...99%, which meets the agrotechnical requirements of 100±20% [3]. Consequently, the dressed seeds can be applied for further use for their intended purpose.

Further in Table 3.13 the results obtained during laboratory and production studies of uniformity of distribution of working fluid over the seed mass are given.

The experimental results are presented in Tables 3.12, and 3.13.

Table 3.12.

Results of laboratory and production tests of dressing completeness

Workin g flow rate liquids, l/t	Seed weight up to mordant ing, g	Seed weight after mordanti ng, g	Average value of seed weight, g		Actual consum ption workin g liquids in convers ion per 1 ton, l	Comple teness of dressin g, %
			before mordanti ng, g	after mordanti ng,g		
1	2	3	4	5	6	7
25		63,128	61,15	62,71	25,51	85
	61,250	62,756				
	61,730	62,798				
	60,910	62,358				
	61,358	62,913				
	60,905	62,776				
	61,216	62,605				
	61,045	62,768				
	61,208	62,552				
	60,992	62,446				
	60,886					
27,5	60,935	62,465	61,1	62,77	27,33	91
	61,258	62,718				
	61,530	63,250				
	61,150	63,000				
	60,805	62,305				
	61,110	62,820				
	61,076	62,506				
	61,233	62,593				
	60,911	62,981				
	60,992	63,062				

30	61,085 61,408 61,680 61,300 60,955 61,260 61,226 61,383 61,061 61,142	62,815 63,162 63,228 63,081 63,240 63,123 63,210 62,916 62,795 63,130	61,25	63,07	29,71	99

As can be seen from the experimental data, the completeness of dressing at the studied (from 25 to 30 l/t) suspension flow rates on the experimental dressing machine is 85...99%, which meets the agrotechnical requirements of 100±20% [3]. Consequently, the dressed seeds can be applied for further use for their intended purpose.

Further in Table 3.13 the results obtained during laboratory and production studies of uniformity of distribution of working fluid over the seed mass are given.

Table 3.13

Results of laboratory and production studies of uniformity of distribution of working fluid over the seed weight

Consumption of working fluid, l/t		25	27,5	30
Seed weight after dressing, g	1	63,128	62,465	62,815
	2	62,756	62,718	63,162
	3	62,798	63,250	63,228
	4	62,358	63,000	63,081
	5	62,913	62,305	63,240
	6	62,776	62,820	63,123
	7	62,605	62,506	63,210
	8	62,768	62,593	62,916

	9	62,552	62,981	62,795
	10	62,446	63,062	63,130
Average value of seed weight, g		62,71	62,77	63,07
Mean square deviation, S		0,23	0,30	0,17
Coefficient of variation, V %		0,37	0,48	0,27
Average error $S_{\bar{x}}$, g		0,07	0,09	0,05

The results of the analysis of the obtained data show that the coefficients of variation of the weight of dressed seeds taken from different places have not high value. Consequently, it can be stated that the uniformity of distribution of working liquid in the seed mass is uniform and the quality of dressing is rather high.

3.9 Conduct a multifactorial study to determine the optimal parameters of the dressing agent with the device of correlation of the rate of suspension consumption to the productivity of the seed metering unit.

After appropriate improvement of the dressing agent, preliminary experiments were conducted to determine the main technological parameters that ensure the treatment of sowing seeds. In the experimental works seeds of breeding variety Namangan-77 with residual pubescence of 8.6% were used.

The criteria for evaluation of cotton seed dressing quality are chosen as the flow rate of suspension U_1 and completeness of dressing U_2 . The main factors influencing the mentioned criteria are: distance from the axis of the swinging tray to the automatic valve L, angle of inclination of the swinging tray to the horizontal plane α, seed dressing productivity P.

According to the results of preliminary studies the levels and steps of variation of factors influencing the quality of dressing were selected (Table 3.14)

When conducting experimental studies, the multifactorial experiment plan B_3 was vibraimed. Matrix of plan B_3 and the results of experiments to determine the quality of seed dressing and processing of experimental data results are given in Appendix 3.

As a result of experimental data processing using computer software, the following regression equations were obtained, which adequately describe the dressing process:

Y_1 =27,168+2,420X_1 -2,533X_2 +1,150X_3 +0,647X_1^2 -0,529X_1 X_2 - 0,546X_1 X +$_3$

+1.047X_2^2 +0.487X_2 X ;$_3$

Y_2 =90,506+8,127X_1 -8,437X_2 +3,830X_3 +2,237X_1^2 -1,775X X_{12} - 1,842X_1 X +$_3$

+3,454X_2^2 +1,583X X ;$_{23}$

Table 3.14

Levels of factors and their variation intervals

№	Factors	Unit	Description of factors		Variation intervals	Levels of variation		
			Natu-ral	Coded		-1	0	+1
1	Distance from the axis of the swing tray to the automatic valve	mm	L	X_1	4	44	48	52

2	Angle of inclination of the swing tray to the horizontal plane	degre e	α	X_2	4	32	36	40
3	Seed dressing capacity	kg/ho ur	P	X_3	500	300 0	350 0	4000

When solving the question of optimizing the parameters of the dressing machine, the following conditions are accepted:

At_1 - the slurry consumption per 1 ton of downed seeds should be no more than 27.5 liters;

$_2$ - Atthe full dressing must be at least 91.5%.

$$AT_1\ (X_1, X, X_{2\ 3}) < 27.5$$

$$AT_2\ (X_1, X, X_{2\ 3}) > 91.5$$

$$X_1 \geq 0, X_2 \geq 0, X\ _3 \geq 0$$

The parameters were optimized using modern computer programs with the help of random search methods. As a result, the following optimal process parameters were obtained:

Factor values	X_1	X_2	X_3
Coded	-1	-0,2324	+1
Natural	44	35,560	4000
Rounded	44	36	4000

According to the results of multifactor studies, we accept the rational value - the distance from the axis of the pumping tray to the automatic valve equal to 44 mm, the angle of inclination of the pumping tray to the horizontal plane equal to 36^0 at dressing productivity equal to 4000 kg/hour.

3.10 Conclusions

1. As a result of experiments it was determined that the distance from the teeth of the saws of the saw cylinder to the regulating wall equal to S = 20 mm and at the protrusion of the saws of the saw cylinder from the grate comb by 7-10 mm, the productivity of the doser is not less than 4000 kg/hour.

2. Experimental studies proved that the developed hydraulic pressure stabilizer provides a more uniform supply of working slurry regardless of the amount of slurry in the flow tank. When working without hydraulic pressure stabilizer the flow rate of slurry when increasing the amount of slurry in the flow tank varies in a wide range and leads to excessive consumption of working slurry and poor-quality dressing of cotton seeds.

3. it is determined that at passage of seeds through the swinging tray there are oscillations at which its angle relative to the horizontal plane changes. From the condition of repayment of the oscillation of the tray experimental studies determined the following parameters: rational mass counterweight-1100 grams; distance from the axis of the rocking tray to the place of its installation - 210 mm and the angle of inclination of the lever to the rocking tray - 75 . ^{0}

4. As can be seen from the experimental data, the completeness of dressing at the investigated (from 25 to 30 l/t) suspension flow rates on the experimental dressing machine is 85...99%, which corresponds to the agrotechnical requirements of 100±20%, while the coefficients of variation of the weight of dressed seeds taken from different places has not high value. Consequently, it can be stated that uniformity of distribution of working liquid in the mass of seeds occurs uniformly and the quality of dressing is quite high.

5. According to the results of multifactor studies, we accept the rational value - the distance from the axis of the pumping tray to the automatic valve equal to 44 mm, the angle of inclination of the pumping tray to the horizontal plane equal to 36^0 to supply the required amount of slurry at the doser capacity of 4000 kg/hour.

IV. PRODUCTION TESTS AND ECONOMIC EFFICIENCY OF COTTON SEED DRESSING FOR DOWNY SEEDS

4.1 Carrying out production studies of the seed dressing machine

One of the most important indicators characterizing the seed dressing machine is its productivity for seeds providing technical requirements established according to the existing regulations on seed preparation. Earlier in theoretical and laboratory researches the productivity of the seed dresser was determined depending on the protrusion of the saws of the saw cylinder from the grate and the distance from the teeth of the saws of the saw cylinder to the regulating wall.

Besides, it was proved that the developed device-stabilizer provides passage through the automatic valve of the necessary amount of the consumed working suspension depending on the productivity of seed dressing.

To confirm the above mentioned were conducted production studies of seed dressing in production conditions. The research was conducted in the seed preparation shop of Kushkupir cotton ginning plant of Khorezm region.

For this purpose, in the subsidiary enterprise of DP "RIM Ustakhonasi" an experimental sample of dressing machine was produced according to the parameters justified by the author, the scheme and general view of which are shown in Figures 4.1 and 4.2.

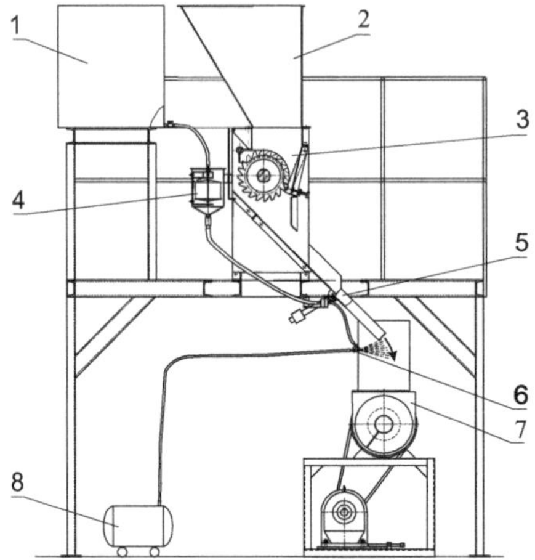

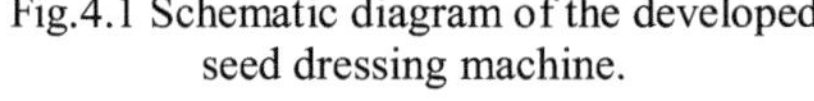

Fig.4.1 Schematic diagram of the developed seed dressing machine.

Fig.4.2 General view of the seed dressing machine.

The experimental dressing machine was installed in the dressing area of the seed preparation shop after the metering hopper. The technological process of dressing on the experimental sample is as follows: the sowing seeds prepared for dressing will go into the upper hopper of the dressing machine. After filling the hopper with seeds, the motor-reducer of the metering unit was started, the speed of the toothed cylinder was set to 60 rpm, other necessary values of parameters determined in laboratory conditions were set (Fig. 4.3).

Fig.4.3 Production research of the experimental sample of seed dressing machine for sowing seeds

The seeds of cotton Mehnat prepared for dressing with 8.5 % pubescence and 3.5 % mechanical damage were used in the research.

Studies were conducted on different productivity of dressed seeds to determine the efficiency of automatic suspension ventilation.

Determination of the actual productivity of the cotton seed dressing machine was carried out according to the method described in the third section. The results of experiments are summarized in Table 4.1 (Annex 4).

Table 4.1.

Determination of the actual flow rate of the working suspension of the dressing agent depending on the capacity of the seed metering unit.

Varin-ty of experiments	Production capacity, t/hour	Flow rate of the working suspension of the dressing agent, l/hour			Average value, l/hour	Actual slurry flow rate, t/l	Increase in mechanical damage of seeds, %
		Repetition of experiments					
		1	2	3			
1	2,5	69,0	68,7	67,8	68,5	27,4	0,9
2	3,0	83,5	83,2	82,1	82,9	27,6	1,0
3	3,5	97,2	96,1	95,3	96,2	27,5	0,9
4	4,0	109,7	110,2	107,6	109,1	27,2	0,9

From the results of Table 4.1 we can conclude that the experimental seed dressing machine provides the required productivity of dressing, the necessary consumption of working suspension per hour of operation and for each ton of dressed seeds.

In addition, the developed automatic valve by means of the rocking chute system provides automatic adjustment of the flow rate of the required amount of working slurry depending on the capacity of the seed dispenser, which is not provided by such adjustment on the existing seed dressing machines.

The results of laboratory analysis of dressed seeds showed that by all indicators dressed seeds meet the requirements of the existing standard for dressed seeds, the completeness of dressing is not less than 80%, the productivity of dressed seeds is not less than 4.0 t/hour, the consumption of working suspension of dressing within 27.5 liters / t and the increase in mechanical damage of dressed seeds is not more than 1.5%.

The analysis of the presented graphical dependencies obtained by theoretical means, laboratory experiments and studies in production conditions show the adequacy of theoretical and experimental studies.

After conducting production research and eliminating design flaws identified during the study, the developed experimental dressing machine of downed cotton seeds was introduced into the technological line for preparation of downed cotton seeds of the cotton ginning plant Kushkupir in Khorezm region. The experimental dressing of downed cotton seeds worked smoothly in the season of 2017-2018, according to the results of which the act of introduction was issued (Annex 5).

Validity of the conducted researches and reliability of implementation of the results of the present researches are confirmed by certificates of superior organizations (appendixes 6 and 7).

4.2. Feasibility study of the equipment complex for dressing of downy cotton seeds

In accordance with the Methodology for calculating the economic efficiency of the introduction of new equipment and production organization for enterprises of the cotton ginning industry, the determination of the expected annual economic effect is based on the comparison of variable costs of the basic and new equipment and is made according to the formula:

$$E = P + [(C_1 + E_н\, TO_1) - (C_2 + E_н\, TO_2)]; (4.1)$$

where: C_1 , C_2 - operating costs by variable cost items, ths.UZS;

P - difference in cost from saving of P-4 preparation consumption (P_1 - P_2), thousand soums;

K_1 ;K_2 - capital investments in basic and implemented variants, ths.UZS;

$E_н$ - normative coefficient of capital investment efficiency (0.15).

Table 4.2 summarizes the input data for calculating the economic effect.

Table 4.2.

Input data for calculation of economic efficiency

№ n/a	Name of indicators	Unit measurements	Basic version	Implemented version
1	Performance	t/h	4,0	4,0
2	Cost of equipment	thousand soums	38400,0	23800,0
3	Installed capacity	kW	4,0	4,25
4	Productive working time	hour	625,0	625,0
6	Electricity charges: -for 1 kWh of energy consumed	sum	450	450
7	Output per year	т	2500,0	2500,0
8	Cost of 1 liter: of the drug (P-4)	sum	33000	33000
9	Drug consumption per 1 ton of seeds	л	4,5	4,125

The cotton seed dressing machine of I-JS-8/L brand was taken as a base variant.

Calculation of the cost of electricity consumption

Electricity consumption is determined based on the power of electric motors installed on the equipment and productive working time

$$W = P_y K_c T_o$$

whereP_y - installed power of electric motors, kW;

T_o - productive time of equipment operation, hour

Energy consumption in the base case:

$$W_c = 4 \text{ x} 625 = 2500 \text{ kWh}$$

Energy cost: $E_{иб}$ =2500 x 450 =1125 thousand soums.

Energy consumption in the implemented variant: W_c =4,25 x 625 =2656 kW/hour

Energy cost: $E_{ив}$ =2656 x 450 =1195 thousand soums.

Calculation of depreciation charges

In the basic version

38400,0 thousand soums x 0,15 = 5760 thousand soums

In the implemented version

23800,0 thousand soums x 0,15 = 3570 thousand soums

Repair expenses

In the basic version

38400,0 thousand soums x 0,05 = 1920 thousand soums

In the implemented version

23800,0 thousand soums x 0,05 = 1190 thousand soums

Table 4.3.

Operating costs

Indicators	Unit.	Options	
		base	project
Amortization	thousand soums	5760	3570
Repair	thousand soums	1920	1190
Electric power	thousand soums	1125	1195

Total	thousand soums	8805	5955

When using the proposed new equipment the consumption of P-4 preparation is reduced by 15% at the rate of consumption of pesticide for 1 ton of seeds 4,5 liters.

Calculating the cost of preperative P-4

Base

$$P_1 = 4.5 \times 33000 \times 2500 = 371250 \text{ thousand soums}$$

project

$$P_2 = (4.5 - (2.5\times0.15))\times33000 \times 2500 = 340312 \text{ thousand soums}$$

Where, 33000 - cost of 1 liter of preparation P-4, sum.

4000 - volume of sown seeds, t.

Difference

$$P = 371250- 340312= 30938 \text{ thousand soums}$$

Table 4.4.

Summary of costs by options

Indicators	Unit.	Options	
		base	project
Cost of equipment	thousand soums	38400,0	23800,0
Operating costs	thousand soums	8805	5955
Cost of the drug P-4	thousand soums	371250	340312
Savings on drug consumption (P)	thousand soums	-	30938

Substituting the obtained values into the formula, we determine the annual economic effect per unit

$$E=[P +(C +E_{БН} K_{Б})-(C +E_{ВН} K_{В})] =30938+ [(8805+0.15\times38400)-(5955+0.15\times23800)]=30938+[14565- 9525] = 35978 \text{ thsum.}$$

Thus, the expected annual economic effect from the introduction of a set of equipment for cotton seed dressing will be approximately **35978** thousand soums.

MAIN CONCLUSIONS OF THE WORK

1. The quality of cotton seed dressing depends on many different factors. The analysis shows that the factors determining the quality of dressing of agricultural seeds can be grouped into four main groups, physical and mechanical properties of seeds, physical and chemical properties of dressing agent, technological factors and factors depending on the design of the dressing agent.

2. Sprayers used in practice mostly disperse the working liquid with the preparation into drops of various sizes, i.e. they form polydisperse systems. In the best case it is possible to regulate the average particle size, and large weight fractions of different-sized particles reduce the effectiveness of treatment and adversely affect the uniformity and area of coverage of the seed surface with the dressing agent.

3. Many types of technologies and technical means for seed dressing of agricultural crops have been developed and introduced into production. However, until now there are no perfect solutions to improve the completeness of dressing, i.e. the compatibility of the amount of dressing agent feeding from the productivity of dressed seeds.

4. Analytical dependences for calculating the values of the distance of the tooth drum protrusion from the grate of the seed dispenser are obtained. Regularities of changes in the trajectory of cotton seeds movement in the zone between the toothed drum and the inclined guide from the change of the toothed drum rotation frequency and the angle of seed ejection in the initial zone are constructed. To ensure the productivity of seed dressing within (4,0÷4,5) tons it is recommended the value of tooth protrusion from the grate within $(7,0\div10,0)\cdot10^{-3}$ m.

5. The graphical dependences of the change of the tray oscillation range on the total weight of seeds in the dressing tray have been

constructed. The analysis of the obtained graphical dependencies shows that with rising of sunrise $\sum m_c$ the tray oscillation range increases according to non-linear regularity. When the total mass of seeds increases from 0.4 kg to 2.4 kg, the tray oscillation range increases from 1.21^0 to 3.09^0 at the mass of 1.5 kg. The recommended values are: load weight 1.3 kg at an output of 4.0-4.5 t/h.

6. Graphical dependences of the change of seed velocity in the tray on the change of the angle of inclination and on the value of seed displacement are plotted. To increase the seed feeding to the suspension zone it is necessary to increase the tilt angle of the tray or decrease the friction coefficient between the seed and the tray surface. The recommended values of parameters are: $\beta=40^0 \div 45^0$, $f=0{,}30$, at which the productivity of seed dresser is provided within (4,0÷4,5) tons.

7. A mathematical model describing small oscillations of the tray with seeds of the seed dresser is obtained. By numerical solution of the problem the regularities of vibration changes of the seed dressing tray are obtained. It is revealed that the amplitude and frequency of the tray oscillations decrease with the increase of the weight mass on the lever. At $\mathrm{m_{гр}} = 1{,}3$ кг, $\sum \mathrm{m_c} = 1{,}08$ кг the range of oscillations of the tray reaches 3.1^0 and, accordingly, the oscillation frequency decreases to 0.71 s.

8. The graphical regularities of change of the oscillation range of the seed dressing tray from the change of the lever angle with the load at variation of the load weight values are constructed. At weight of a cargo 1,6 kg increase of angle β from 25^0 to 50^0 leads to increase of the oscillation range of the tray with seeds from $0{,}51^0$ to $2{,}03^0$, and with decrease of the mass of the load $\Delta\zeta$ varies from 1.23 to 5.67^0 . Recommended values of the angle of the lever with the load in relation to the seed tray are ($75^0 \div 80$).^{0}

9. As a result of experiments it was determined that the distance from the saw teeth of the saw cylinder of the saw cylinder of the seed doser to its regulating wall equal to S=20 mm and at the protrusion of the saws of the saw cylinder from the grate comb on 7-10 mm, the productivity of the doser is not less than 4000 kg/hour.

10. Experimental studies proved that the developed hydraulic pressure stabilizer provides a more uniform supply of working slurry regardless of the amount of slurry in the flow tank. When working without hydraulic pressure stabilizer the flow rate of slurry when increasing the amount of slurry in the flow tank varies in a wide range and leads to excessive consumption of working slurry and poor quality dressing of cotton seeds.

11. As can be seen from the experimental data, the completeness of dressing at the studied (from 25 to 30 l/t) suspension flow rates on the experimental dressing machine is 85...99%, which corresponds to agrotechnical requirements of 100±20%, while the coefficients of variation of the weight of dressed seeds taken from different places has not high value. Consequently, it can be stated that the uniformity of distribution of working liquid in the mass of seeds occurs uniformly and the quality of dressing is quite high.

12. According to the results of multifactor studies, we take a rational value - the distance from the axis of the pumping tray to the automatic valve equal to 44 mm, the angle of inclination of the pumping tray to the horizontal plane equal to 36^0 to supply the required amount of slurry at a doser capacity of 4000 kg/hour.

13. The results of production tests of the recommended dressing agent have shown that by all indicators dressed seeds meet the requirements of the existing standard for dressed seeds, the completeness of dressing is not less than 80-85%, the productivity of dressed seeds is

not less than 4.0 t/hour, the consumption of working suspension of the dressing agent within 25-30 liters / t and the increase in mechanical damage of dressed seeds is not more than 1.5%.

14. Expected annual economic effect from introduction of recommended cotton seed dressing agent will be tentatively **35978** thousand soums as of 2019.

LIST OF REFERENCES

1. International cotton advisory committee. Washington, DC, From the Secretariat of the ICAC. https://icac.org/, email secretariat@icac.org. October 7, 2019.

2. Presidential Decree No. UP 5708 of April 17, 2019 "On measures to improve the system of state management in the field of agriculture" the main tasks of the Ministry of Agriculture of the Republic of Uzbekistan.

2. To the Resolution of the Cabinet of Ministers of the Republic of Uzbekistan No. 604 of December 23, 2004 "On measures on improvement of organization of cotton seed production".

3. Kuchkarov H.M., Akramov A.A., Abdugabborov S. PDI 54 - 2015 Uruglik chigitni dorilash bўyicha tavsiyanom.

4. Andreeva E.I., Pronchenko T.S. Cotton seed dressing agents. Materials of the meeting on pre-sowing preparation of cotton seeds. Tashkent, 1978. C. 10-12.

5. Khoshimova A. Modern state of cotton growing abroad (review), Tashkent, 1976. 52 c.

6. Khoshimova A. Modern state of cotton growing abroad (review), Tashkent, 1978. 38 c.

7. https://propozitsiya.com/osobennosti-normirovaniya-protraviteley-semyan-zernovyh-kultur

8. Deryabin V., Irgashev E. Dredging of cotton seeds. //Agriculture of Uzbekistan. - 1967. - №1. - C. 8-10.

9. Dzhasheev A. Seed quality and accuracy of seedling distribution over the feeding area // Tractors and agricultural machinery. - 2004. - №1. - C. 37-39.

10. Drincha V.M., Tsydendorzhiev B., Kubeev E.I. Basic principles of pre-sowing chemical dressing and physical disinfection of seeds // Agrarny Expert. -2009.

11. New in preparation of sowing cotton seeds. /Expressinformation. - Tashkent, UzNIINTI. - 1987. - 6 c.

12. New cotton seed dressing agent //Information sheet of UzNIINTI. - Tashkent. - 1988.

13. Modern means of mechanization for pesticide application and prospects of their development // Journal of Mendeleev All-Union Chemical Society. - 1984. - T. 29. - №1

14. https://agrovesti.net/lib/tech/plant-protection-tech/obrabotka-semennogo-materiala-protravlivanie-semyan.html

15. Pavlov I.F. Protection of field crops from pests / - 2nd edition, supplement. and revision. - Moscow : Rosselkhozizzdat, 1987. - 256 c.

16. https://www.syngenta.kz/rekomendacii-po-tehnologii-obrabotki-semennogo-materiala

17. Jamaliev A. Black root rot // Mechanization and electrification of agriculture. 1979. №9. C. 14-16.

18. Tagirova V. et al. Estimation of cotton infestation by black root rot // Cotton growing. 1982. №4 c. 24.

19. https://agromage.com/stat_id.php?id=158

20. https://www.cropscience.bayer.ru/kornievyie-ghnili

21.https://cyberleninka.ru/article/n/kornevye-gnili-zernovyh/viewer Ovsyankina A.V.

22. https://ru-ecology.info/term/59243/

23. http://www.cawater-info.net/library/rus/iwrm/iwrm13.pdf

24. Mukhin V.D. Dredging of seeds of agricultural crops. M., Kolos, 1971. 46 c.

25. Bakhramov K.B. Instruction on pre-sowing preparation of downy cotton seeds in farms. Tashkent. 1986. 14 c.

26. Polityko P. M. Protection of winter crops in autumn // Plant Protection and Quarantine. - 2008. - № 8. - C. 20-22.

27. Semynina T. B. Sow only dressed seeds // Plant protection and quarantine. - 2008. - № 8. - C. 43.

28. Bugaev P. Good seeds - good harvest // Rural mechanizer. - 2005. - № 5. - C. 26-27.

29. Dorozov A. Influence of pre-sowing seed treatment with Pectin and microelements on the quality of winter wheat, pea and soybean yield // Grain farming. - Moscow : 2001. - № 1 (4). - C. 31-33.

30. Kalashnikov K.Ya. Combating dusty blight of wheat / - Moscow Leningrad : Selkhozizdat, 1955. - 64 c.

31. Lepekhin N.S. What gives seed dressing // Plant Protection. - 1994. - № 3. - C. 39.

32. Methodical instructions on dressing of seeds of agricultural crops. - Moscow : Kolos, 1984. - 48 c.

33. Niyazov A.M. Pre-sowing treatment of barley seeds in the electrostatic field : Cand. Candidate of Technical Sciences : 05.20.02 / Izhevsk GSA. - Izhevsk, 2001. - 125 c.

34. Talanov I.P. Influence of stimulating preparations on yield and quality of wheat grain // Grain farming. - 2001. - № 4 (7). - C. 21-22.

35. Teplyakov B.I. Factors of increasing productivity of spring wheat // Plant Protection and Quarantine. - 2004. - № 4. - C. 24-25.

36. Tishkin V. T. The role of plant protection is obvious // Plant Protection and Quarantine. - 2009. - № 12. - C. 4-6.

37. Klavins U. V. Treatment of tubers before planting // Potatoes and Vegetables. - 1977. - № 5. - C. 12-13.

38. Zabazny P. A., Buryakov Y. P. Brief reference book agronomist / - Moscow : Kolos, 1983. - 320 c.

39. http://www.cawater-info.net/library/rus/iwrm/iwrm13.pdf

40. https://www.activestudy.info/gommoz-xlopchatnika/

41. Drincha V. M., Kubeev E. I. Basic principles of pre-sowing chemical dressing and physical disinfection of seeds // Agrarny Expert. - 2009. - № 3.

42. https://ogorodstvo.com/bolezni-rasteniy/bolezni-khlopchatnika/sistema-meropriyatij-zashhity-xlopchatnika-ot-boleznej.html

43. Baiguskarov M.H. Perfection of the drum etchant for pre-sowing seed treatment with biopreparations : Cand. Sci. (Techn.) : 05.20.01 / [Place of defense: Bashkir State Agrarian University] - Ufa, 2011. 143 c.

44. Dzhuraev R.H. Justification of technology and basic parameters of the device for dressing downy seeds of cotton during dragging. dissertation. candidate of technical sciences. Yangiyul, 2000. 136 c.

45. Gabrakhmanov I.H., Nurullin E.G., Erov Y.V., Salakhiev D.Z. et al. Recommendations to ensure the quality of bread harvesting, post-harvest processing of grain and seeds of cereals, legumes and cereal crops/- Kazan: Publishing and Printing Company, 2011.

46. Gabdrakhmanov I.H., Erov Y.V., Karimov K.Z., Kuzmina T.I., Yashin D.A. // Reference book on seed production of cereals, leguminous and cereal crops - Kazan, 2009.

47. Dolzhenko V.I., Kotikova G.Sh., Zdrozhevskaya S.D. et al. Seed dressing / - M., 2003. - 62 c.

48. Dorogov E. Test-drive of dressing agents // New Agrarian Journal.- March-May 2011. - №2(2).

49. Semynina T.V. Sow only dressed seeds. // Plant protection and quarantine. - 2008. - №8. C.43.

50. Tenyaev A.V., Doiskova N.M. Good seed - good sprouts // Plant Protection and Quarantine. - 2004. - №3. C.12-13.

51. Trishkin D.S. Handbook of agronomist on seed dressing of grain crops. Recommendations for quality dressing / Edited by Dr. D.S. Trishkin - M., 2006.

52. Volovik A.S., Glez V.M., Zamataev A.I. et al. Potato protection from diseases, pests and weeds: reference book /- Moscow : Agropromizdat, 1989. - 205 c.

53. Reference book on primary processing of cotton.Tashkent-2019, 268-271 p.

54. Salakhov I.M. Development and substantiation of parameters of pneumomechanical seed dressing of grain crops. Cand. Sci. Tech. Dissertation: Kazan-2014.

55. Klenin N.I., Sakun V.A. Agricultural and land reclamation machines: Elements of the theory of working processes, calculation of adjustment parameters and operating modes. 2nd ed., rev. and supplement. - M.: Kolos, 1980. - 671 c.

56. Listopad G.E., Demidov G.K., Zonov B.D. and others. Agricultural and meliorative machines - M.: Agropromizdat, 1986. - 688 c.

57. Myslyvchenko A.V., Lavrentiev S.P., Usatykh N.A. Technology of mechanized agricultural work: methodology for practical training. Part IV: Means of mechanization for plant protection from pests and diseases / compiled by. Novosibirsk State Agrarian University: Engineering Institute. - Novosibirsk, 2007. - 27c.

58. Karpenko A.N., Khalansky V.M. Agricultural machines. - 5th ed., rev. and supplement. - M.: Kolos, 1983. - 495 c.

59. Technology of mechanized agricultural works. Methodical instructions for practical classes. - Novosibirsk, 2007.

60. https://www.agroinvestor.ru/technologies/article/15125-da-budet-dozhd/

61. https://www.agrodialog.com.ua/protravlivanie-semyan.html

62. Vasiliev V. Seed dressing and harvest // Grain crops. - Moscow, 1993. - № 1. - C. 20-21.

63. Egorycheva M. T. Effectiveness of pre-sowing seed dressing // Plant Protection and Quarantine. - 2009 - № 8. - C. 43-44.

64. Zhitkov S. P. Preparation of seeds for sowing /- Leningrad : Lenizdat 1958. - 74 c.

65. Kalashnikov K. Ya. Seed dressing / - Moscow : 1961. - 62 c.

66. Chemical and biological plant protection / edited by P. A. Khizhnyak. - Moscow : Kolos, 1971. - 215 c.

67. Peresypkin V.F. Diseases of industrial crops. M., Agropromizdat, 1986. 312 c.

68. Rakhmanin V.G. Research of processes of treatment and dressing of seeds with application of electrostatic field. Dissertation of Candidate of Technical Sciences. Chelyabinsk, 1974. 186 c.

69. http://www.cnshb.ru/AKDiL/0048/base/RR/050010.shtm

70. Shmonin V.A., Drichik S.T. Improvement of pesticide and fertilizer spraying technology // Tractors and agricultural machines. - 2000. - № 5. - C. 34-36.

71. Dityakin Yu.F., L.A.Klyachko, B.V.Novikov, V.I.Yagodkin. Atomization of liquids. - Moscow : Mashinostroenie, 1977. - C. 208.

72. https://cyberleninka.ru/article/n/raspylivanie-zhidkostiforsunkami/viewer

73. Dunsky V.F., Nikitin N.V., Sokolov M.S. Pesticide aerosols / - Moscow : Nauka, 1982. - 288 c. 96

74. Vladykin, I.R. Control of the installation for seed pre-sowing treatment by UV radiation (in Russian) // Mechanization and electrification of agriculture. - 2007. - № 10. - C. 8.

75. Gavrilova O.P., T.Y. Gagkaeva. Fusarium of grain in the north of the Non-Chernozem region and in the Kaliningrad region in 2007-2008 // Plant Protection and Quarantine. - 2010. - № 2. - C. 23-25.

76. Dunsky V. F., Nikitin N. V. V., Sokolov M. S. Monodisperse aerosols / - Moscow : Nauka, 1975. - 188 c.

77. Khasanov E.R. Analysis of fine particle crushing by disk atomizers // Problems of agroindustrial complex in the South Urals and Volga region : proceedings of the regional scientific-practical conference of young scientists and specialists.-Ufa : BSAU, 1999.-S. 162-167.

78. Dunskiy V.F., Nikitin N.. V. Liquid atomization by a rotating disk and the question of "secondary" droplet crushing // Engineering Physics Journal. - 1965. - T. 9, № 1. - C. 54-60.

79. Gavrilov, A. A. Biopreparations for winter wheat protection against diseases // Plant Protection and Quarantine. - 2001. - № 1. - C. 29.

80. Rakipov V.G., Akramov A.A., Jamolov R.K., Abdughabbarov S.A. "Dorilangan chigit etalonini tayirlash kurilmasini islab chikish va rationale volchamlarini aniqlash" // "Fan, ta'alim v islab chikarish integrationlashuvi sharoitida innovatsionnogo tekhnologii dolzarb muammolari" maususidagi Respubliki ilmiy-amaliy anjumani. Toshkent-2017. C.22-25.

81. Emelin B.N. Investigation of a new technology of seed dressing: Cand. Sci. (Techn.) Dissertation: 052001. - Saratov, 1970. - 23 c.

82. Maslo I.P. and others. Mechanization of protection of plants. - K.: Urozhay, 1982. - 144 c.

83. Azizkhodjaev U.H., Dyachkov V.V., Rakipov V.G. Study and testing of the equipment complex for dressing cotton seeds with the preparation "Furadan" of the American firm FMS and preparation of proposals. RESEARCH AND DEVELOPMENT WORK. Tashkent 1988.

84. Pokras V.M., Dyachkov V.V., Rakipov V.G., Daminov G.D. Determination of the possibility of using the FMS plant for cotton seed dressing with preparations of Soviet production. RESEARCH WORK. Central Research Institute of Cotton Industry, Tashkent 1991.

85. Pokras V.M., Normukhamedov T.A. Processing of seed cotton at cotton cleaning plants (review).-Tashkent, 1977.- pp.22-29.

86. http://k-a-t.ru/sxt/3-zashita2_protrav/index.shtml

87. Sigayev E.A., Altergott A.A., Shadrin S.A., Poluyanov A.Y. Seed dressing machine. Russian Federation patent No. 2217897.2003.

88. Maslov G.G., Kozhan V.N., Mechkalo A.L., Barisova S.M. Seed dressing agent. Patent of the Russian Federation №2316925. Bulletin No.5.2008.

89. Sabirov K.S., Rakipov V.G., Jamolov R.K. Development of a universal dressing for downy and bare cotton seeds. NTO. Paxta tozalash IICHB, Tashkent. 2006. 7-9 c.

90. Usmonov V.U. "Researches on improvement of cotton seed preparation technology", Avtoref. Diss.k.t.n., T.1976, 14-15 p.

91. Rakipov V.G. et al. Development of hopper-doser of downy seeded cotton seeds. STO. JSC NPC "Paxtasanoat ilm", Tashkent. 1999. 92 c.

92. Rashragovich A.Y., Korabelnikov R.V., Bordanovsky V.V. and Rashragovich Y.A. Feeder of fiber processing machine. USSR patent No. 1379359.Bul.No.9.1988.

93. Dyachkov V.V., Maksudov E.T., Rakipov V.G. et al. Bunker-doser of downy cotton seeds. Patent IAP 02654.Bul.No.2.2005.

94. Jamolov R.K., Akramov A.A. et al. Development of a dressing machine with a device for correlation of the rate of suspension consumption according to the performance of the seed doser, to improve the efficiency of dressing.-Tashkent, 2012.-29-42 p. (Report/JSC "Paxtasanoat ilmiy markazi"). (Report/Paxtasanoat ilmiy markazi PJSC)

95. Djamolov R.K., Akramov A.A. "Dori suyukligining mejorii sarfini chigit doser unumdorligiga moslashtiruvchi kurilma"// Tukimachilik muammolari. №1 2014 й. Б.15-19.

96. Akramov A.A., Jamolov R.K. "Technology of cotton seed dressing and equipment for its implementation"// Collection of scientific articles IV International Scientific and Practical Conference June 04-05. Kursk-2014. C. 27-29.

97. Uz FAP Patent No. 00873. "Urug dorilagichi" Kushakeev B.Y., Gulyaev R.A., Jamolov R.K., Tuychiev V.H., Akramov A // Rasmii akhborotnoma.-2014.

98. Uz FAP Patent No. 01412. "Ishchi suspension bosimini bosimini stableashtiruvchi qurilma" Khozhiev MT, Akramov AA, Jamolov R.K., Nazirov R.R., Kuchkarov H.M // Rasmii akhborotnoma.-2019.

99. Akramov A.A., Jamolov R.K. "Analysis of the Trajectories of Seeds Falling on an Inclined Guide in the Etching Machine"// International Journal of Advanced Research in Science, Engineering and Technology. Vol. 7, Issue 4, April 2020., http://www.ijarset.com/upload/2020/april/30-paxtailm-march-15.pdf

100. Djuraev A.D., Daliyev Sh.L.. Development of the dasign and justification of the parameters of the composite flail drum of a cotton cleaner.// European Sciences review Scientific journal No. 7-8 2017, p.96-100.

101. Maksudov R.H., Dzhuraev A. Machina va mekhanizmilar nazariyasi 2 qism (Machina va mekhanizmilar dynamikasi). Ўқuv ккўllanma ISBN 978-9943-11-946-8 "Fan va tekhnologii" nashriyoti, Toshkent 2019. 300 б.

102. Jamolov R.K., Akramov A.A., Dzhuraev A., Turdiev H.. "Analysis of small oscillations of tray mordanting plant sowing downy seeds of cotton"// FarPi ilmiy-tekhnika journal. Vol. 24. No. 4. 2020 й. Б.139

103. Augambaev M., Ivanov A.Z., Terekhov Y.I. "Fundamentals of planning a research experiment". T., Ukituvchi. 1993 г.

104. Akramov A.A. "Improved treater for the pubescent planting seeds" // 76th Plenary Meeting of the International Cotton Advisory Committee (ICAC) "Cotton in the of globalization and technological progress" XIII International Uzbek cotton and textile fair Tashkent-2017 page.-87-90.

105. Khozhiev M.T., Akramov A.A., Jamolov R.K., Ubaidullaev M.M. // Determination of the rational parameter of dressing machine with a device for correlation of the rate of suspension consumption respectively with the performance of the seed doser. Collection of scientific articles of the International Scientific and Practical Conference December 22-23, 2016. VOLUME 2. Kursk 2016. 343-347 c.

Printed by Books on Demand GmbH, Norderstedt / Germany